To Dwell
with a
Boundless
Heart

Studies in the
Postmodern Theory of Education

Joe L. Kincheloe and Shirley R. Steinberg
General Editors

Vol. 77

PETER LANG
New York • Washington, D.C./Baltimore • Boston
Bern • Frankfurt am Main • Berlin • Vienna • Paris

David W. Jardine

To Dwell with a Boundless Heart

Essays in Curriculum Theory, Hermeneutics, and the Ecological Imagination

PETER LANG
New York • Washington, D.C./Baltimore • Boston
Bern • Frankfurt am Main • Berlin • Vienna • Paris

Library of Congress Cataloging-in-Publication Data

Jardine, David William.
To dwell with a boundless heart: essays in curriculum theory, hermeneutics,
and the ecological imagination / David W. Jardine.
p. cm. — (Counterpoints; vol. 77)
Includes bibliographical references.
1. Education—Philosophy. 2. Education—Curricula—Philosophy.
3. Ecology. 4. Hermeneutics. I. Title. II. Series: Counterpoints
(New York, N.Y.); vol. 77.
LB14.7.J37 370'.1—dc21 97-32372
ISBN 0-8204-3941-X
ISSN 1058-1634

Die Deutsche Bibliothek-CIP-Einheitsaufnahme

Jardine, David W.:
To dwell with a boundless heart: essays in curriculum
theory, hermeneutics, and the ecological imagination / David W. Jardine.
–New York; Washington, D.C./Baltimore; Boston; Bern;
Frankfurt am Main; Berlin; Vienna; Paris: Lang.
(Counterpoints; Vol. 77)
ISBN 0-8204-3941-X

Cover design by Nona Reuter.

The paper in this book meets the guidelines for permanence and durability
of the Committee on Production Guidelines for Book Longevity
of the Council of Library Resources.

© 1998 Peter Lang Publishing, Inc., New York

Printed in the United States of America.

TABLE OF CONTENTS

ACKNOWLEDGEMENTS

These essays have been reprinted here because of the generous permission of the editors of the journals involved. What follows are the original citations for these works:

Jardine, David W. (1990) On the humility of mathematical language. *Educational Theory*. 40: 181–92.

Jardine, David W. (1990) Awakening from Descartes's nightmare: On the love of ambiguity in phenomenological approaches to education. *Studies in Philosophy and Education*. 10: 211–32.

Jardine, David W. (1990) "To dwell with a boundless heart": On the integrated curriculum and the recovery of the Earth. *Journal of Curriculum and Supervision*. 5: 107–19.

Jardine, David W. (1992). Immanuel Kant, Jean Piaget, and the rage for order: Ecological hints of the colonial spirit in pedagogy. *Educational Philosophy and Theory*. 23: 28–43.

Jardine, David W. (1992) "The fecundity of the individual case": Considerations of the pedagogic heart of interpretive work. *Journal of Philosophy of Education*. 26: 51–61.

Jardine, David W. (1993). Wild hearts, silent traces, and the journeys of lament. *Journal of Educational Thought*. 27: 16–20.

Jardine, David W. (1993) "A bell ringing in the empty sky." *JCT: An Interdisciplinary Journal of Curriculum Studies*. 10: 17–37.

Jardine, David W. (1994) Student teaching, interpretation, and the monstrous child. *Journal of Philosophy of Education.* 28: 17–24.

INTRODUCTION

In re-reading the essays for this collection, I was struck by how single-mindedly they are in working and re-working a nest of interrelated ideas.

In these essays, I have explored and cultivated the affinities between hermeneutics, curriculum and ecology. I believe that each of these disciplines is at its best when organized around ideas of interrelatedness, generativity, relationships of (shared and contested) kinship and ancestry, and the cultivation of whole and healthy relationships between the young, new, or innovative and the old, established, and proven. I believe, therefore, that each of these disciplines can learn from the language and spirit of the others and that each can take the lessons learned to heart.

In my work in undergraduate education, elementary-school practicum supervision, and in my own research, I have found this growing sense of affinity to be quite a relief. It disburdens education from having to remain within the weightiness, humorlessness, and literalism of psychological discourse about children, about our lives with them, and about the disciplines with which we have been entrusted as educators. It means that there is not such a great difference, for example, between teaching a graduate course in "Hermeneutic Inquiry" and teaching an undergraduate course in "Methods in Early Childhood Education." Both involve cultivating in oneself the ability and desire to keep the world open. This simply means that good teachers, like good "interpreters," must cultivate in themselves and the children they teach a sense of the *interpretability of the world.* The world of mathematics, for example, is not closed and fixed and finished. It is a living and breathing discipline, full of rich relations and recursions of meaning, full of histories and ancestors and deeply buried implications. And this interdependent, interrelated, ongoing living character of the whole of mathematics is present *right here,* in this child's rhythmic clapping, in these measured steps, in these brow furrows over fractions and their ways.

Certainly cracking open the discipline of mathematics beyond the flat surfaces of black-line masters makes our lives and the lives of our children more difficult, more ambiguous, more ongoing. However, such work is also more pleasurable, more engaging, more sustaining, more nourishing, and more real than schooling often allows. Knowing that the world is full of living disciplines and that there is real, intimate, good-hearted work to be done, is good news for us all.

In this way, an interest in the old, established, and the proven need not lead to branding hermeneutics "conservative." A hermeneutic interest in tradition and ancestry (an interest in what Gary Snyder called "the old ways") requires not simply the protective repetition of such traditions. Hermeneutics incites the particularities and intimacies of our lives to call these traditions to account, compelling them to bear witness to the lives we are living. Hermeneutics demands of such disciplines and traditions that they tell us what they know about keeping the world open and enticing and alive and inviting. And, to the extent that such disciplines and traditions can no longer serve this deeply pedagogical purpose, to that extent they are no longer telling, no longer helpful in our living, no longer true.

Thus, hermeneutics sits squarely on the same cusp as education itself: the roiling space between the established and the new, between the young and the old. A sometimes dangerous, sometimes funny spot.

What is also evident in these essays is a slowly emerging reconsideration of the nature and necessities of writing. Some of these essays contain rather hard-nosed, rather arcane, and difficult explorations of complex philosophical issues that underwrite education (Aristotelian logic, Cartesianism, ideas of univocity and mathematical self-identity, Kantian philosophy and how it underwrites Piagetian theory, Biblical quarrels about analogical discourse, and so on). Detailed, too, are the philosophical ins and outs of the philosophical origins of phenomenology, its lingering legacy of essentialism and the recovery of hermeneutics out from under this legacy via the later writings of Ludwig Wittgenstein. Any way you take it, this is a rather imposing list of names and issues that seem far removed from the everyday realities of the classroom. However, consider the following examples.

Unlike many semi-literate "qualitative research methods" textbooks in education which suggest that phenomenological research is about "people's experiences," a careful examination of Edmund Husserl's work shows that all experiences are experiences *of* something and that this "something" (that *of which* you or I have experiences) is the "topic" of a phenomenological exploration. Such a profound insight into the "topographical," worldly nature of human experience provides a way to avoid the horrible excesses of "teacher narratives" and the heady claims to ownership and critical immunity in phrases such as "this is my experience" or "this is my story." It also avoids the tendency for interpretive work to degenerate into a sort of psychologistic mire of "subject's experiences" and their representation in "themes" gleaned from transcripts. In addition, carefully portioning off the tendency towards essentialism in Husserlian phenomenology will provide a way to avoid the equally horrible excesses of a moralizing phenomenological pedagogy which is a bit too sure of itself and the essences of things. Without intentionality, we become isolated into a sort of Democratic Cartesianism, with each experiencing subject locked in a self-empowered bubble, everyone with something to say but no one living out any obligation to listen. Without a critique of essentialism, it becomes impossible to understand how our children might have a say in what we take to be so essential. In neither world is pedagogy possible, let alone necessary to being human.

Finally, a brief word about ecology. This, perhaps, has been my greatest relief and greatest pleasure: envisaging education as a deeply Earthly task, part and parcel of the breath of the world. Ecology offers both education and hermeneutics images of the old and the young, of places and histories, of disciplines and work, that break out beyond the confines of the human voice, out into the pitter-patter of an animate, living Earth, *of which* the disciplines of schooling are a part, not an exception. Thinking about curricular disciplines as open fields of living relations that not only undergo but require constant renewal and transformation makes these disciplines inviting again, full of the sparkle of old wisdoms and the sparking voice of the child. Out of relation to each other, both the old and the young become monstrous hallucinations—ecological, intellectual, and spiritual disasters waiting to happen.

CHAPTER ONE

Awakening from Descartes's Nightmare: On the Love of Ambiguity in Phenomenological Approaches to Education (1990)

Prelude

After recently observing a student-teacher complete a lesson with a group of about twenty seven-year-old children and listening to her accounts of how the lesson went, I mentioned that she seemed to be "somewhere else" during the lesson, appearing vaguely unhappy, not "there" somehow. I suggested that this "disconnectedness" might have been partially responsible for why the children seemed to stray—they seemed to be getting the message that things were not alright and were reacting appropriately to the deep message she was giving that underlay her rather forced surface articulations of enthusiasm. Her responses were rather frightening, but also vaguely sad and almost poignantly humorous: "You mean you want me to smile more?", "Maybe I should have used more eye contact or something?"

We began a long and painful conversation about the fact that she was speaking about smiling or looking at children as if they were just tricks to use; we talked about the way that such things must come from the heart, that they are living relationships that had to inform who she *was* with children; we stumbled our way through her initial talk of "having a bad day," and "not being planned enough" and began to face the fact that this incident was rather typical of the past weeks of her practicum. She didn't seem to want to "be there" with them; she seemed to be missing that difficult and ambiguous attunement and genuineness to which young children are naturally drawn.

Admittedly, such attunement to children and the ability and desire to face them in a living way is a difficult thing to judge in another person. Such matters are especially not amenable to being on a "checklist of student-teacher competencies." I can more easily determine, for example, the frequency of the use of eye contact than I can determine whether such eye contact is genuine. And I can more easily define "genuine" as "effective" because this will allow the specification of particular consequent behaviors of the child in terms of which "genuineness-as-effectiveness" can be (more) clearly determined, more easily "checked." But this still sidesteps the problem that it could all be for effect. And, as we have come to say without qualm in some circles, what difference does it really make, as long as there is the effect? We can never judge "genuineness" in another, or caring for children or having a deep interest in their lives with any certainty and therefore, judging the competence of a student-teacher who may come to teach my six-year-old son Eric cannot be done on such a basis. Matters such as genuineness, care, love, patience, integrity, trust, listening, attunement, a respect for the deep difference of children—matters which, as every teacher and parent knows, lie at the heart of living our lives with children *even though* such matters are difficult, ambiguous, and risk-laden—are shunted into the realm of "uncheckable" and out of the realm of criteria for teacher evaluation. Moreover, mentioning my own son in an academic paper becomes understandable only as a form of sentimentalism; educational inquiry often recoils from the point at which they might be about *someone,* preferring irresponsible objectivity (or, better, objectivity that is responsible only to the methods of its production) to a responsive voice.

In the teacher-education courses in which I am involved, it is much easier to pass on to students the latest research which documents correlations between frequencies of eye contact and measures of teacher effectiveness, than it is to reflect and converse with them on the interweaving meanings and experiences involved in facing someone in a genuine, pedagogic way. When we begin to do such reflection, two phenomena arise. First of all, there is no definitive and declarative endpoint to such reflections. This phenomena *remains* a difficult and vibrant feature of being a teacher of young children, something we return to over and over again in the practice of teaching, reconsidering it, re-thinking it, facing it anew with each new child, each new day. Gaining a living understanding means living with this uncertainty and fluidity. Secondly, we can not be "disconnected" from the topic of our reflections: even though genuinely facing a child is a phenomenon we must *all* confront as teachers, how *I myself* am connected to and implicated in this phenomenon inevitably arises. Suddenly, in the midst of attempting to understand a common feature of teaching, I am no longer just anyone. Who I am and how I live my life with the children I teach is implicated in this reflection. Self-understanding and self-reflection are required from which no theory will exempt me. My patience, my frustrations, my tolerances and preferences, my deeply held beliefs as to whether children are worth listening to, whether this child, here, now, is one for whom I care—all of this issues forth in the living experience of facing children and thereby, issues forth in my attempts to understand this lived experience. Such reflections voice a deep connectedness, a deep interestedness (*inter esse*—being in the middle of things), a deep investment in the issues at hand: it is *theorizing* in the best sense, a theorizing that erupts out of our lives together and is about our lives together.

When the living character of education is rendered by a desire for clarity and distinctness, all that remains is what can be objectively documented without difficulty and ambiguity. All that remains is that of which I can talk with indifference, disinterestedness, anonymity, and the other hallmarks of objectivity. Reflection on the experience of facing children in a pedagogic way is foregone in favor of that which can be known with some certainty, with little risk—genuinely facing children becomes replaced with frequencies of eye-contact episodes. Genuine and deep attention to some matter becomes managerialized into acceptable time-on-task ratios. In all of this, we render children into strange and silent objects which require of us only management, manipulation, and objective information and (ac)countability. Children are no longer our kin, our kind; teaching is no longer an act of "kindness" and generosity bespeaking a deep connectedness with children. In the name of clarity, repeatability, accountability, such connections become severed in favor of pristine, "objective" surface articulations. And "reconnecting" with children must proceed under the auspices of such clarity—"smiling," for example, becomes a technique to be applied because research has demonstrated its effectiveness. Instead of a heartfelt and difficult and, in many ways, unresolving conversation in response to my student's question "You mean you want me to smile more?", the professional response has become, for some in education, simply "Yes." It may be little wonder that some teachers in whose classrooms we place student-teachers accuse us in academia of not really *knowing* what we are talking about.

Introduction

How could I deny that these hands and this body are mine, were it not perhaps that I compare myself to certain persons, devoid of any sense, whose cerebella are so troubled and clouded by the violent vapours of black bile, that they constantly assure us that they think they are kings when they are really quite poor, or who imagine that they have an earthenware head or are nothing but pumpkins or are made of glass. At the same time, I must remember that I am . . . in the habit of sleeping and in my dreams representing to myself the same things or sometimes even less probable things, than those who are insane in their waking moments. (Descartes, c.1640, 1955, p. 90)

I resolved to assume that everything that ever entered my mind was no more true than the illusions of my dreams. But immediately afterwards I noticed that whilst I thus wished to think all things false, it was absolutely essential that the 'I' who thought this should be somewhat, and remarking that this truth, *'I think therefore I am'* was so certain and assured that all the most extravagant suppositions brought forward by the sceptics were incapable of shaking it. And then, examining attentively that which I was, I saw that I could conceive that I had no body, and that there was no world nor place where I might be; but yet that I could not for all that conceive that I was not. (p. 29)

The prelude of this paper is not offered as an isolated episode. It voices deep currents that haunt our lives as educators and that may be coming to haunt the lives of our children. We are still living in Descartes's dream, but this dream is slowly, perhaps inextricably, becoming a nightmare. In education, we must consider whether we are passing on this nightmare to our children—the nightmare of pristine self-clarity and dominion over the Earth as paradigms of understanding, self-understanding, and mutual understanding—or whether we are opening up the possibility of awakening from it.

Descartes's dream was that the questioning steps of the then newly emerging sciences could adhere to the first principle of thought—"I think therefore I am"—and could participate in its indubitability. Descartes's work had intended to free inquiry from unquestioned obedience to authority, but it did so at what is slowly turning out to be a nightmarish cost. All of the moist and dark and ambiguous connections of our lives on Earth, to each other and to the lives of our children, had to be *severed* and could be reachieved only to the extent that such connections moved within the parameters of the clarity and distinctness won by his methodical doubt of everything that could be doubted. Descartes took the first steps towards making the Earth suffer the tyranny of a subject able to contact anything outside of itself only within the methodical parameters of its own self-presence and self-security. He took the first steps towards, on the one hand, bringing forth the inevitability of the subject as a moment of inquiry (ushering in the age of method as a way of containing the acts of the subject—they must be methodical and reproducible) and, on the other, disconnecting and estranging that subject from life as it is actually lived. This disconnecting, coupled later with the ways in which the then-emerging human sciences came to ape the natural sciences in the nineteenth century, not only estranged the subject from life as it is actually lived, but rendered our lives knowable only *through* such disconnectedness. In order to understand life as it is actually lived, we must disconnect ourselves from it and then reconnect with it only in those ways that render it our predictable and manageable object. It is under this shadow that we can speak, now, in an unruffled way, of providing life-management courses in high school and have come to slowly transform being a parent into parenting skills, being a teacher into teaching skills. It is under this shadow that research into classroom life has often begun with the assumption that teacher and children in the classroom have nothing especially interesting to offer inquiry *precisely because they are a living part of that classroom,* deeply connected with and invested in it.

One of the claims of this paper is that we are living out this disconnectedness and disengagement in the sphere of educational theory and practice in often unnoticeable and unvoiced ways. We are silently living out Descartes's dream-turned-nightmare. As we sever our connections with the Earth, it ceases to be our abode and becomes a meaningless objective mechanism which is at the disposal of our whim and consumptive fantasies. And, correlatively, as the Earth loses its *humus,* its living, generative character, the subject loses its humanity by losing the connectedness with the humus out of which it has emerged. One might say that the subject loses its humility, its Being-in-the-world, its sense of having a place on Earth, and becomes, in Descartes's nightmare, a disembodied and worldless self-presence to which the Earth must submit. Our lives, the lives of our children, the life of the Earth, become well lit, enlightened, presentable, clear, univocal, like thin veneer whose surface is unambiguous, shiny/reflective, clean, without depth—bodiless, sexless, ghostly, empty, shallow products of what Alfred North Whitehead called the "celibacy of the intellect" (cited in Fox 1983, 24).

The full contors of what has become Descartes's nightmare are too complex to address fully in the present context. They involve images of ourselves and the Earth that are rooted deeply in the North American spirit, images the questioning of which may prove not only difficult and painful, but also not widely desirable. In this paper, I wish simply to trace a line of thought aimed at unearthing an underlying impetus of educational inquiry that comes from this dream: the tendency in some areas of educational theory and practice has been towards specification, univocity, clarification, and, essentially, the overcoming of ambiguity. The deep ambiguities of life as it is actually lived, the deep difficulties in living our lives with children are often designated, either explicitly or implicitly, as the enemy of discourse and therefore as the enemy of true understanding. These difficulties and ambiguities are understood as *problems to be fixed,* things to be "cleared up" through the diligent pursuit of research which takes as its first gesture a fundamental severance with its object of inquiry so that it can heed only its own desire for clarity and distinctness which then *demands* clarity and distinctness from that object.

Such a desire will not be found listed as the explicit *intent* of any particular inquiry. It operates, rather, as an unvoiced presumption that goes without saying in educational theory. At bottom, life is presumed to be essentially unambiguous, clear and distinct—a possible *object* of research which can be disassembled and clarified, within statistically documentable parameters, piece by relentless piece—and finding it ambiguous indicates a problem that simply has not yet been fully addressed, fully researched, and fully specified in all its inherent clarity.

It is around this presumption that a great divide originates in the field of educational theory. Those involved in forms of inquiry that have their origins in phenomenology seem to have a peculiar tolerance for ambiguity—perhaps, even stronger, a love of ambiguity. And it is precisely such love of ambiguity that makes phenomenology and its related disciplines intolerable to those outside this tradition, for it (often peevishly) refuses Descartes's dream of clarity and distinctness, preferring instead to attempt to voice the contours of life as it is actually lived, facing up not only to the difficulty of such a life, but to the insecurity involved in finding such a voice. John Caputo (1987) goes as far as to "define" phenomenological hermeneutics, not as a method for clarifying or solving questions regarding some feature of life, but as a "restoring of life to its original difficulty" (1), to the contours of the *Lebenswelt,* the lived world, everyday life. Phenomenology and its related disciplines hold open the promise, or at least openly express the desire, to disrupt Descartes's dream-turned-nightmare.

The first section of this paper is an attempt to find trace-lines of Descartes's nightmare in a deep correlation of most forms of educational inquiry—the correlation between the *univocity of reality* and *univocal discourse.* This correlation cashes out as an intolerance for ambiguity in inquiry. It is an attempt to reproduce the lives of children, the life of the classroom, the *curriculum vitae,* into clear and presentable objects borne out of a severance from life as it is actually lived.

This is followed by a brief exposition of the core of the work of Edmund Husserl, the "father" of contemporary phenomenology. This core is the *phenomenological reduction,* which involves putting out of play the deep, assumed correlate of univocity that underlies most forms of inquiry. It is this putting out of play that accounts for the often frustrating *difference* of phenomenology. It is this putting out of play that signals the turning of the attention of phenomenological inquiry to life as it is actually lived, with all the ambiguity that such living entails.

This section is followed by a brief Interlude, in which the old extremes available to inquiry are shown to haunt Husserl's work. Analogical discourse is proposed as an alternative to Husserl's options of Heraclitean flux on the one hand, and fixed essences on the other hand. Analogical language, following Ludwig Wittgenstein (1968), is proposed as a way of giving a voice to life as it is actually lived, with all the ambiguous "kinships" (36) or "family resemblances" (32) that such a life involves.

The final section of this paper is a brief reflection on Martin Heidegger's formulation of thinking and inquiry as forms of obedience and thanksgiving. This notion of Heidegger's is not pursued as a form of romanticism, nor is it offered as a quaint curio. Rather, it is offered as an attempt to respond to an all too easily forgotten phenomenon in education. We are in the presence, as educators, of new life in our midst. "The basic fact of education is natality: the fact that children are *born* into the world" (Arendt 1969, 174). We stand, in education, at the moment of the generativity of the human race and education is, most essentially, our response to this moment. But it cannot be denied that, given the deeply consumptive desires of North American culture, given the ecological horrors left in the wake of Descartes's nightmare, we may be standing at the moment of the *degenerativity* of humanity. Something is afoot, something is called for in these times that goes beyond "one more damn thing," (Smith 1988a) one more theory to be dumped on the pile with others in our onrush to ecological self-consumption.

This image of inquiry as obedience and thanksgiving raises the questions of our place on the Earth and offers a moment of quiet in the din of "the latest, the newest, and the best in educational theory and practice." Heidegger's later work does not offer what Maurice Merleau-Ponty (1970a) called a nostalgia for "our relationships to Being such as they were prior to self-consciousness," prior to Descartes's dream, since we "owe [our] idea of and our taste for primordial ontology to just this self-consciousness. There are some ideas which make it impossible for us to return to a time prior to their existence" (154). Descartes's bringing forth of the subject as a moment of inquiry is precisely such an insurmountable idea.

Rather than nostalgia, perhaps what is required is what Thomas Berry (1988) called a "post-critical naivete" (5)—a naturalness or simplicity of speech that disrupts our burgeoning edifices of knowledge which are threatening to collapse under their own weight, threatening to do nothing more than exhaust us in their nightmarish hurry to finally get things right. This is the naive, perhaps impossible hope of phenomenology—to turn us away from our idealized and admittedly beautiful and seductive edifications and grand theories, and back to life as it is actually lived. It is a perhaps impossible hope that we can recover our humility, our humanity, our humus, our living place on the Earth. All around us is the urgency. For the ecological consequences of believing in our own dominion are accelerating and threaten to suddenly trivialize all our earnest theorizing and demonstrate to us full force that we are not worldless, self-present subjects who can live in the rarified atmosphere of Descartes's dream.

ReDreaming the Nightmare: The Univocity of Reality, Univocal Discourse, and the Obedience to Method

As mentioned above, there is no particular inquiry to point to in addressing Descartes's nightmarish legacy. And the threads of this legacy extend far before Descartes's own work:

> For not to have one meaning is to have no meaning and if words have no meaning, our reasoning with one another, and indeed with ourselves has been annihilated; for it is impossible to think of anything if we do not think of one thing; but if this is possible, one name might be assigned to the thing (Aristotle, 3rd C. B.C.E.).

"It is impossible to think of anything if we do not think of one thing": this is an expression of the principle of non-contradiction and its correlative, the principle of identity—A=A. It expresses a simple, basic, and natural assumption we make about the world and objects in the world. Things are, so to speak, identical with themselves. That is to say, when we speak of an entity, or attribute something to it, that entity either *is* or *is not* what is so attributed. However unclear, ambiguous, or blurred our understanding, experience, or language may be, entities in the world clearly, unambiguously, and precisely *are* what they *are*. "The essence of truth is identity" (Heidegger 1928, 1978, 39) and, "taken as a negative expression of the principle of identity, the principle of non-contradiction is the basic principle of all cognition, of all truths as identities" (52). Even under the auspices of Immanuel Kant's Copernican Revolution, identity persists as a criterion of knowledge; reason demands that the object of knowledge be self-identical. In the work of Kant (1767, 1964, 106), and later in the genetic epistemology of Jean Piaget (1952, 9), identity is a category of knowledge.

Knowledge cannot contain ambiguity or unclarity. And the paradigm of clarity is just this: A=A.

Since a basic assumption of all cognition is that things are the way they are (what could be called the "univocity of reality," A=A), to speak about such things, our speech must reproduce such singularity. To bespeak the self-sameness or univocity of reality, speech itself must be univocal. Since "it is impossible to think of anything if we do not think of one thing" (A=A), "one name might be assigned to the thing" in order to give a voice to this self-identical nature. Irrespective of whether we "locate" this self-identical nature in "things themselves" to which thought must submit, or in the essence of cognition (which then "constructs" objects according to its own demand for self-identity), "identity is the basic criterion of all truth" and therefore a basic criterion of all true discourse. To speak truly is to adhere to the univocal frontiers of things themselves (or to adhere to the univocal frontiers of reason itself which reproduces its own frontiers by constructing objects of knowledge in light of those frontiers). Discourse must revolve around "one name."

Here we have a basic correlation which provides a basis for inquiry: the deep-seated belief in the univocity of reality requires that speech *about* such univocal entities must itself be univocal. In the face of this, the arising of ambiguity, equivocity, metaphorical speech, analogy, rhetoric, poetry—all of these are taken as quite literally contradictions (i.e., peculiarities of speech which do not reflect the nature of things themselves). In order to properly signify this nature, we find that "the notion of signification requires a univocity of meaning: the definition of the principle of identity, in its logical and ontological form, demands it. Univocity of meaning is ultimately grounded in essence, one and self-identical" (Ricouer 1970, 23). To speak truly about things is to submit speech to the univocal character of the thing itself. (Or, under the auspices of Kant's Copernican Revolution, to speak truly is to submit experience to the univocal core of Reason itself.) We can hear in this echoes of the long history of the notion of "substance" as that which stands by itself and rests within its own frontiers, that "which requires nothing but itself in order to exist" (Descartes c.1640, 1955, 275).

The next question that immediately arises is this: how are we to gain access to such a univocal character? It becomes necessary to assure the adherence of speech to the univocal character of things themselves. Since "a single truth alone is acceptable when we are dealing with a problem of knowledge in the strict sense" (Piaget 1965, 1971a, 216–17), signification which makes claims to be true must orient to univocity, to "one voice." We must secure ourselves in those techniques or methods which will promote such a singularity of voice (this being what Habermas [1972] named the "monological" character of scientific discourse). Discourse becomes formalized (MacDonald 1975, 283), orienting to precision, definition, and repeatability, orienting, in a sense, to "form" or "essence," to that which is self-identical. The desire to define becomes warrantable as a correlate of the definite character of the things which are to be designated by such definitions. The methodical purging of discourse of its ambiguities operates under the desire to match up discourse with the unambiguous character of things themselves. This is why formal logic tends to form the backdrop of such discourse because, as Jean Piaget demonstrates, such discourse perfectly adheres to this deep, presumptive correlation between discourse and things themselves. Formal logic is perfectly "equilibrated" since the structures of thought/language in terms of which we do logic and the object of such doing *are precisely the same.* Formal logic "proceeds by the application of perfectly explicit rules, these rules being, of course, the very ones that define the structure under consideration" (Piaget 1968, 1970, 15). In formal logic, the frontiers of discourse and the frontiers of the object of discourse are identical. The lines that have been drawn are identical, because the matters under consideration have become matters of method (i.e., matters regarding the ideal, formalized *operation* of discourse itself).

In some quarters of educational inquiry, the matters of deepest concern become the methods whereby some educational issue is rendered clear, precise, and defined. Method gains a primacy to the extent that it begins to determine which matters are warrantable objects of investigation. (Reflection on the experience of facing a child in a genuinely pedagogical way becomes unwarrantable, because it cannot be subjected to the clarity, precision, and definition demanded by certain warranted methods. It is thus subjected to what has become a commonplace banishment— since such reflection does not result in something clear and distinct, it is not "objective," and if it is not "objective," it must be simply "subjective"). We have come upon an age where the *matters* of inquiry have become matters of method (Heidegger 1962, 1972, 55 ff.), since it is only method that will save us from simply talking about ourselves and remaining enslaved in subjectivity. Perhaps this is why the "higher" stages of the development of knowledge in Piagetian theory are defined by their *operational* character. The highest level of the development of knowledge, for Piaget, is logico-mathematical knowledge, i.e., knowledge of the operation of knowledge itself, in short, a knowledge of matters of method. Objectivity is defined as adherence to such methods, since it is precisely such methods which *constitute* an object of knowledge in the first place. Through a peculiar convolution, then, Piaget's notion of structures, defined as systems of transformations of operations, comes hauntingly close to Descartes's scholastic notion of substance as that which need nothing other than itself in order to exist. His structuralism operates under "an ideal (perhaps a hope) of intrinsic intelligibility supported by the postulate that structures are self-sufficient and that, to grasp them, we do not have to make reference to all sorts of extraneous elements" (1970, 4–5).

Connected with this turn to method, and perhaps equally inextricable in our image of inquiry, is the "relentless passion for quantity" (Merton 1972, 2) as a paradigm of self-identical clarity and distinctness. In discussing Gottfrid Leibniz's notion of adequate knowledge, Martin Heidegger (1928, 1978) states:

> An adequate knowledge is thoroughly clear knowledge, where confusion is no longer possible, where the reduction into marks and moments of marks (*requisita*) can be managed to the end. Of course, Leibniz immediately adds regarding *cognitio adaequata:* "I am not sure that a perfect example of this can be given by man, but our notion of numbers approximates it." (62)

The paradigm of univocity and identity is *numerical identity* and its various statistical permutations. (It is no coincidence that the dreamer of Descartes's dream was most deeply and passionately a mathematician). Number— quantification—becomes paradigmatic to the extent that even unclarity and indistinctness become quantified into designations of the possible frequency of error in one's results.

This reliance on numeration is no stranger to education: "the reduction into marks . . . can be managed to the end" signals how old is our impulse to link up knowledge of a child with those things about which we can be clear, ("where confusion is no longer possible"), with the number of clearly specifiable correct answers on a test, with numbers, with marks, frequencies and distributions. Just to complete the circle, we also pass on this reliance to our children: they sometimes don't "make the grade." Sometimes they "fail," with a finality that is much clearer and more distinct than the reasons that can be given for it.

This turn towards mathematization is done under the presumption that, no matter how unclear we may be, *the matter itself is clear,* and thereby, whatever offers us such clarity is presumed ahead of time to give us the matter at hand. (Despite our difficulty in understanding a child in our classroom, they unequivocally *have* certain abilities, and therefore the unequivocal "mark" they get on a test takes on the appearance of corresponding to such unequivocal abilities). Mathematization is, again, the deep desire to get re-connected with things, to understand them. Our lives with children might *appear* ambiguous, but, in fact, they are not ambiguous, but simply not as yet fully clarified in their self-identicalness. We may voice a sort of mathematical hesitancy regarding the possibility of full clarity, but such hesitancy refers not to the thing itself, but to our *methods of knowing* and to the degrees of confidence and volume with which we can declare such knowing. After all, since such hesitancy refers to the possibility of *error,* such a reference entails the presumption of a self-identical entity *from which* we (may) have deviated in a statistically documentable way. Without such a presumed identity, there can be no identification, no final and definitive saying of what this or that thing *is,* no definition. Even though it can be demonstrated that a margin of error is always necessary in such quantifiable approaches (and therefore perfect clarity and distinctness are not possible "in reality"), perfect clarity and distinctness must remain *conceivable* for such approaches to remain meaningful in their designations of *margins* of *error.* Although it is never possible in reality, "the last word in education" (i.e., the final full clarification of some educational phenomenon, such that nothing more can be or needs to be said about it) remains something deeply desirable and potent. We long for a sense, in some educational circles, of an *objectivity* that is pronounceable with the gusto and confidence and clarity of the "I am." We long to be *right.*

The Phenomenological Shuffle

Such a longing may also be heard as a deep lament for a sense of connectedness that has been lost, a lament for a relation to the Earth, to our lives, to the lives of children. It may not be so much that we long for objectivity. It may be that we long to *be* right. It is this lament, this sense of profound crisis and dis-ease, that propelled Edmund Husserl's phenomenology. He maintained that we cannot *make* this connectedness by simply moving ahead more diligently with the articulations of Descartes's dream. A traumatic "break," a "rupture" is required:

> If we miss the meaning of the phenomenological reduction, everything is lost. The temptation to misunderstand is almost irresistible (Husserl c.1925, 1960, 163).
>
> Perhaps it will even become manifest that the total phenomenological attitude and the epoche belonging to it are destined to effect . . . a complete personal transformation, comparable . . . to a religious conversion, which then, however, over and above this, bears within itself the significance of the greatest existential transformation which is assigned as a task to [humanity] as such (Husserl 1932–34, 1970, 137).

These two passages vividly portray the zealousness and fervor of phenomenology at its best and worst. There is something irremediably *different* about how phenomenology begins, proceeds, and how it conceives the ends of inquiry. Phenomenology centers around "experience" and it is obsessive about how inquiry can be obedient to experience in its fullest sense, how it can turn away from the edifices that our theorizing has built up, and turn back to "things themselves," just as they give themselves to be in our experience. This is why Husserl, the father of contemporary phenomenology, found it necessary to introduce and re-introduce phenomenology over the course of forty years. It is also why Maurice Merleau-Ponty was able to ask, in the preface to his seminal *Phenomenology of Perception,* "What is phenomenology? It may seem strange that this question has still to be asked half a century after the first works of Husserl. The fact remains that it has by no means been answered" (1970b, vii). In a sense, however, it is no longer strange that now, nearly 90 years after the publication of Husserl's *Logical Investigations,* the question "what is phenomenology?"—the question, essentially, of what *difference* it makes—remains as potent and peculiar as ever. This question retains its potency because, once the difference of phenomenology is understood, it not only puts into question how inquiry is usually conducted, but it flies in the face of some of the deepest hopes and aspirations of contemporary North American culture. It is a different way of understanding ourselves and our place in the world, one which problematizes our aspirations to clarity, progress, mastery, and dominance as images of our relation to the Earth and to each other. It brings inquiry out from under the desire for the final Word; it opens us up for the rebirth and re-enlivening of the Word in the soul, with the full richness and ambiguity that such re-enlivening requires.

It is thus not trivially but rather deeply subversive, wanting to give a voice to the living text and texture of human life that underlies our idealisms, our objectifications, and our plentiful fantasies. It wants to return life to itself, to re-awaken us to life as it is actually lived. *This,* for Husserl, is the *real* foundation of inquiry: everyday life. Phenomenology, then, is not just one more framework or model or method for understanding the world that can be listed alongside a plethora of other approaches, *as if* it lived in the same desire as other forms of inquiry—to render the world and our Being-in-the-world clear, distinct and unambiguous. This is not to say that phenomenology has no desire. Rather, its desire is not redemptive. It does not want to redeem everyday life through the application of methods that will render it presentable according to some imagined norm of clarity and distinctness. Its desire is not to render our experience of the world, but to give a voice to it just as it is.

As Husserl maintains, "phenomenological explication does nothing but *explicate the sense this world has for us prior to any philosophizing . . . a sense which philosophy can uncover, but never alter"* (1930, 1970b, 151). The difference of phenomenology's interest in the sense and experience of the world prior to any philosophizing/theorizing centers around its relation to what Husserl called the "thesis of the natural attitude." It is this attitude to the world that undergirds Descartes's dream, for it implicitly pronounces the deep belief in the univocal character of the world, and, thereby, the univocal character of true speech. It is this attitude which phenomenology "puts out of play" in its own vocation:

> I find continually present and standing over against me the one spatio-temporal fact-world to which I myself belong, as do all other men found in it and related in the same way to it. This fact-world, as the word already tells us, I find to *be out*

there, and also, take it just as it gives itself to me as something that exists out there. All doubting and rejecting of the data of the natural world leaves standing the general thesis of the natural standpoint. The world is as fact-world always there; at the most it is at odd points "other" than I supposed; this or that under names such as "illusion" or "hallucination" and the like must be struck *out of it,* so to speak; but the "it" remains ever, in the sense of the general thesis, a world that has its being out there (Husserl 1913, 1969a, 106).

It is a natural *assumption* about the world that it has its being "out there." That is to say, independent of experience, language, etc., the world is what it is ("the 'it' remains"). We may run into various difficulties in understanding that world—under names such as illusion, hallucination, ambiguity, unclarity, equivocation, and the like. But these difficulties do not affect the fact that the world has its self-identical being out there, and such difficulties must be struck out of discourse if it is to be true to this being out there. Even in such striking out, the "it" remains, ever self-identical, ever calling for univocal discourse to give it a voice.

It is, according to Husserl, precisely this thesis that undergirds Descartes's dream, because it is precisely this thesis that is first *negated* and then *reclaimed* by his methodical doubt. The question persists in Descartes's work—how do I know that these ideas that I have are ideas of things "out there?" How do I make this "move?" The fundamental *problem* in Descartes's meditations becomes one of how the self-present subject can touch something outside of itself, thus keeping in force the desire that undergirds the thesis of the natural attitude by *beginning with* a subject which dreams that it is worldless, out of touch, severed from the Earth by the demands for clarity and distinctness. And he concludes that this subject can do so through precisely the clarity and distinctness of those ideas that participate in such self-presence—these are (*via* God's assurance) ideas *of* the world "out there." It is these ideas that provide such a desired transcendence. Once encapsulated in its own self-presence, only the clarity of that self-presence will do. The thesis of the natural attitude is not disrupted or put out of play in Descartes's nightmare, but simply negated and then regained through its mathematization. Descartes retains the belief in the "being out there" of the world, but simply transforms this sense of "being out there" in light of the *cogito*—to *be* out there is to be clear and distinct. All the ambiguous ways in which things were experienced to be out there cannot withstand his methodical doubt. The living Earth and our lives together withdraw into silence.

It is its relation to this thesis of the natural attitude that distinguishes phenomenology and its related disciplines from other forms of inquiry. It is the fundamental *disruption* of this thesis as an assumption from which inquiry proceeds that gives phenomenology its difference. Phenomenology does not begin by *denying* the principle of identity, nor does it begin by *affirming* this principle. It does not wish to declare what the "being out there" of the world might be, seeking some alternate univocal *foundation* for inquiry to be placed alongside God, substance, sense-data, empirical fact, etc., in the history of philosophizing. It begins by "putting out of play" our tendency to *assume* such a principle as that to which experience must submit in order to be admitted into the boundaries that inquiry has drawn for itself. It puts out of play our tendency, in inquiry, towards foundations, towards causes or explanations or suppositions which will "found" this experience in something not present in the experience itself. This act of disruption, this act of "putting out of play" our tendency to invest experience with transcendent suppositions which go beyond things as they are experienced, is the *phenomenological reduction.*

As cited above, it is important to understand what the reduction does. For if we miss the meaning of the reduction, phenomenology becomes lost in quarrels with skepticism, subjectivism, idealism, and psychologism, *and it is none of these.* It is none of these, not because of peculiar distinguishing features which make its "standpoint" different. It is none of these because it is not a standpoint at all: "we start out from that which *antedates* all standpoints" (Husserl 1913, 1969a, 88). Phenomenology is not some version of how we might make Descartes's dream come true—how we might make a defensible stand in relation to the world and our Being-in-the-world. It is, rather, an attempt to express how things *already stand* with us in the world, how we are *already* right in the middle of things. It wishes to re-enliven the objectifications and boundaries with which we are *already living* (that which "antedates all standpoints" is not some pristine, untouched "experience," but the full complexity of our lives, with all their hidden assumptions, interpretations, and hopes) and win back the surging, delicate life that both informs such boundaries and slips between the cracks, frustrating the desire for *stasis,* foreclosure, and clarity. Phenomenology does not see such fluidity and ambiguity as an error, as a lack of vigilant clarification, but as an indication that we "have continuously anew the *living truth from the living source*" (Husserl 1910–11, 1969b, 279).

In order to free us to give a voice to how we already stand, to voice life as it is actually lived, Husserl formulates the phenomenological reduction as a way of loosening the hold of the thesis of the natural attitude. This is at once a loosening of the hold of our belief in the univocal character of our Being-in-the-world and the subsequent belief that understanding the world requires something special, some special method or procedure to get us in touch with the world. By unearthing the intentionality of experience, the reduction shows that experience is always and already an experience *of* something. We are *already* connected to the Earth, to each other, to our children, albeit in ambiguous and multivocal ways.

The reduction thus sets us into relation with the thesis of the natural attitude such that "we make no change in our convictions" (Husserl 1913, 1969a, 108). We do not go from believing that the world has its being "out there" to denying this thesis. The reduction is not a matter of "a transformation of thesis into antithesis, of positive into negative"(108) but a suspension of all position taking about the being of entities which transcend our experience of them, our life with them—phenomenology wishes to "see" what our place, our life, our lived-experience *is*. We are to faithfully document our experience-of-the-world just as it gives itself to be, without taking a position on that experience or judging that experience in light of some fantastic and seductive paradigm of clarity.

Where does the phenomenological reduction take us, then? Well, it doesn't take us anywhere. It doesn't move us from our actual lives on this Earth to some otherworldly realm; it does not wish to pretend, as science must, that it knows nothing of life until its methods are enacted; it doesn't desire to substitute idealized modes of rationality, discourse, and action for the actual ways in which our lives are conducted. It doesn't take us away from the world outside into the "moist gastric intimacy" (Sartre 1970, 4) of subjectivity. It doesn't provide us with a new standpoint, new methods, a new framework, a different model, an alternate theory, another perspective, a better picture, a clearer view. Things are *exactly* what they were before *except for this one blessed difference*—the desire to redeem the world through the wielding of the weapons of clarity, objectivity, truth, and univocity, has been "let go." It leaves us right where we always already were, with the actual play and interplays of life, with all its difficulty and ambiguity, unredeemed or, better, not in need of redemption but only thoughtful savoring, reflection, conversation, and understanding.

In such letting go, a whole edifice of inquiry begins to come apart at the seams. No longer is the impetus to methodically penetrate the assumed univocal character of things themselves. No longer are we driven in inquiry by the desire to enslave the Earth with images of knowledge-as-dominion. No longer does inquiry in education skip over the head of life as it is actually lived and posit a clear and univocal phantasm world to which life must now live up and of which our experience of life is but an epiphenomenon. No longer is the impetus to secure the methods of inquiry prior to the in flux of experience itself. The securing of such methods *always comes too late.* Or, better, such a securing of methods erupts *out of* life as it is actually lived, as a moment of it, as a set of actions we must live with. We don't require a method to achieve a relationship to and understanding of life—we *are* alive *even if* this life is more confusing, joyous, and difficult than our methods can tolerate. It is this already living understanding of the world, this lived-world, that is of interest to phenomenology.

Thus, when the phenomenological inquirer enters a classroom, he or she can no longer do so with the assurance of exemption, domination, and authority: something is *already* at play, and the living character of this setting is not waiting upon the inquirer for some beneficent bestowal of meaning. It is *already* meaningful, and these connections must not be severed in order to understand them; they must, rather, be delicately gathered in all their contingency, locatedness, and difficulty. Moreover, the inquirer's desire to understand this classroom, this situation, this child, this setting, is no longer something especially distinctive. Understanding this situation is something everyone in this classroom is *already* involved in, teacher and children alike. Inquiry, phenomenologically conceived, can thus not be preemptive, importing exogenous methodological assurances into life as it is lived.

The deep desire for univocity in Descartes's dream does not disappear in phenomenological inquiry. Rather, it reappears out from under its sedimented objectifications—it appears *as* a desire. It is not denied or dispelled as if it were *false,* as if we could simply ignore it at our leisure.

Rather, its full ludic character is brought out as something at play in our lives. The research reports, the teacher-effectiveness checklists, the grades and marks, do not simply disappear. They reappear as features of the life of education, as lived phenomena of education. To inquiry into the full phenomenon of education requires that we face up to our hope for clarity and savor it *as a feature of our lives,* not as indicative of some hidden univocal world that underlies our lives and which our lives constantly fail to be. It requires a critical reflection on the social and cultural "locale" of such hopes for clarity. It requires considering the intimate interrelations that have grown between forms of understanding that offer clarity and distinctness and institutionalized forms of administrative and managerial accountability. It requires considerations of how "the good teacher" has become transformed into enumerable accountability. It requires considering how our students' lives are driven by haunting images of succeeding and getting ahead, how this is linked with marks having become the reward for work done, how making the grade has become an unquestioned function of diligence and effort, and, in all of this, how becoming educated has become a consumable item involving unambiguous checks and balances. ("Tell me exactly what you want us to do on this assignment," "Tell me when to smile, tell me what to say when I reflect on what it is like to face a child in a genuine way.") It requires addressing how organizations like schools have transformed differences into deviations and how finding one's life, one's course(s) difficult is pathologized dis-ease and education is directed towards quelling dis-ease.

Phenomenology lays out for educational inquiry a painful task of articulating our actual lives as educators. But, in a sense, the bad news turns out to be the good news. Phenomenology raises the possibility of real hope, i.e., the hope that life as it is actually lived can be faced. It maintains that we, as educational "theorists," have a living connectedness with the "subjects" of our inquiry. These children in this classroom, this teacher, are not distant objects to which our only relatedness is one forged out of a calm, methodological indifference; they are us, our kind, our kin, and understanding them is understanding our kinship with them, understanding, not severing, the ties that already bind us to the Earth, to our lives, to the lives of our children.

Interlude: Fixed Essences and the Heraclitean Flux

We cannot see how, in the *epoche,* the "Heraclitean flux" of constituting life can be treated descriptively in its individual facticity (Husserl 1932–34, 1970, 177).

Not even the single philosopher by himself, within the *epoche,* can hold fast to anything in this elusively flowing life, repeat it with always the same content, and become so certain of its this-ness and its being-such that he could describe it, document it, so to speak (even for his own person), in definitive statements. But [such] full concrete facticity can be . . . grasped in another good sense, precisely because . . . the great task can and must be undertaken of investigating the . . . fact . . . as belonging to its essence, and it is determinable only *through* its essence (178).

The life-world does have, in all its relative features, a *general structure.* We can attend to it in its generality and, with sufficient care, fix it once and for all in a way equally accessible to all (139).

Husserl's work fell prey to the old options of inquiry, the old extremes—on the one hand, we have the facticity of experience, which, in and of itself, offers us only *differences,* only minutia all speaking with equal, different voices. It offers us only *equivocity.* On the other hand, we have *essences* which can be fixed once and for all. We have *univocity.* Given the task of voicing life as it is actually lived, it is *assumed* by Husserl that this voice must either be the single, isolated voice of difference (such that lived-experience turns out to be idiosyncratic and subjectivistic), or the clear and foreclosing voice of identity (such that the truths of lived-experience turn out to be nestled in an articulation of its general structure, to the extent that some "phenomenological pedagogues" can use the articulation of "essences" as a moral reprimand against the unenlightened, thus reproducing in phenomenology precisely the horrors of Descartes's nightmare, now at the level of "experience" instead of "objects"). What Husserl lacked—what, in fact, he did not even seek, since fixed essences served his desire for philosophy as a Rigorous Science of Being—was a deeper, more difficult voice that refused such a resolution of life into this either/or.

Such a refusal is found in the later work of Ludwig Wittgenstein:

As in spinning a thread, we twist fibre on fibre. And the strength of the thread does not reside in the fact that some one fibre runs through its whole length, but in the overlapping of many fibres (Wittgenstein 1968, 32).

Don't say: "There *must* be something common [some essence, some univocal core of meaning]" but *look and see* whether there is anything that is common to *all.*—For if you look at them you will not see something that is common to *all*, but similarities, relationships, and a whole series of them at that. To repeat: don't think but look! (31)

We see a complicated network of similarities, overlapping and criss-crossing: sometimes overall similarities, sometimes similarities of detail. I can think of no better expression to characterize these similarities than "family resemblances" (32).

These passages portray for me the deeply *dialogical* and *analogical* character of lived-experience, the deeply *conversational* nature of life as it is actually lived, with its irresolvable and potent "family resemblances" and "kinships." They hint at the practical, lived struggle for communication and meaning, which is not solved through univocal definitions and declarations, no matter how clear and distinct. Regarding our desire, in Descartes's dream, to draw univocal boundaries, Wittgenstein rather playfully says "that never troubled you before when you used the word" (1968, 33). Descartes's dream does not trouble the lived usage of language, the lived character of experience:

If someone were to draw a sharp boundary, I could not acknowledge it as the one that I too always wanted to draw, or had drawn in my mind. For I did not want to draw one at all. His concept can then be said to be not *the same* as mine, but akin to it. The kinship is just as undeniable as the difference (36).

This is not the context in which to draw out the full implications of Wittgenstein's work. A brief comment on the nature of analogical language is called for, however, since it does rely on precisely the sorts of fragile interrelationships of meaning that are lost in the glare of Descartes's dream. It does require becoming conversant with one's object of inquiry in such a way that the nightmare of identity and the presentability of literalism and clear definitions is disrupted in favor of the rich kinships of our lives on the Earth.

Analogical thinking involves the exploration of likenesses, similarities, correspondences—parallels between worlds of discourse (*ana logos*). Such parallels resist the collapse of one realm of discourse into another, while also resisting the isolation of such realms into their differences. They involve, so to speak, the "conversation" or "dialogue" between such realms, a dialogue which sustains a "similarity-in-difference" (Norris-Clarke 1976).

Understanding an analogy, therefore, is not a matter of discovering some discursive, univocal term which makes both sides of the analogy *the same,* which collapses the tensive "kinship" it evokes into literal terms that can be applied *univoce* to both sides of the analogy. Analogies do not involve finding some identical "A" in both sides of the analogy and then securing and safeguarding understanding by being able to declare "A=A." The essence of the truth of an analogy is *not* identity (but neither is it difference). It is found, rather, in the lively, propulsive, and difficult *tension between similarity and difference,* a tension that *cannot be cashed out discursively in just so many words.* "Understanding" an analogy is not had through the application of an univocal method, but is had by becoming party to the conversation between different realms of discourse that the analogy opens up, "getting in on the conversation." To "understand" an analogy is not to close it down through some univocal declaration, but to keep it going, to keep it alive, to dwell in the kinships, relations, and similarities it evokes.

> Here it is difficult as it were to keep our heads up—to see that we must stick to the subjects of our every-day thinking, and not go astray and imagine that we have to describe extreme subtleties. We feel as if we had to repair a torn spider's web with our fingers (Wittgenstein 1968, 46).

Our imagination leads us to dreams of clarity that require methodical diligence. Sticking to the subjects of our everyday thinking is sticking, not to such dreams, but to life as it is lived. Attending to difficult and ambiguous kinships or family resemblances of living our lives, as educators," with children all around us, "means living in the belly of a paradox wherein a genuine life together is made possible only in the context of an *ongoing conversation* which never ends and which must be sustained for life together to go on at all. The openness that is required is not a vague licentiousness, but a risky, deliberate engagement of the full conflict and ambiguity by which new horizons of mutual understanding are achieved" (Smith 1988b, 175).

As Wittgenstein documented so well, if we look at language as it is actually used, it is more fluid, flexible, lively, and risk-laden than Descartes's nightmare of clarity would have us imagine. Phenomenologically, this character is not a *mistake* which inquiry must *fix*—believing this is simply a *delusion* (i.e., an attempt to take the play, the interplays, out of language). Wittgenstein's notion of family resemblances also captures the fact that the alternative to clarity, distinctness, and identity is *not* chaos, meaningless facticity, individualistic personal opinion, or idiosyncrasy, but rather is analogical kinship, a belonging together without being identical or simply different. The alternative to Descartes's nightmare of identity is *conversation.*

Before moving on to some concluding remarks, a few "skips" in argument that have been assumed in this paper need to be brought out more explicitly. From Wittgenstein, we have the notion of language as involving analogical relationships, conversations between realms of discourse which resolve into neither identity nor difference. This has implicitly been used as a metaphor for an image of understanding the Earth and our place on the Earth—one, not of the methodical rendering of the Earth in light of Descartes's dream, but of becoming conversant with the Earth in a way that the delicate kinship and interrelationships of and with the Earth can be brought out. Understanding the Earth in an ecologically sound way requires something other than Descartes's narcissistic legacy of dominance and mastery; it requires something other than pristine clarity. Truly understanding the Earth requires a delicacy of discourse that has a kinship with the Earth. The Earth is not simply and straightforwardly our *object.* It is also our *home.* It *sustains* us, *surrounds* us; we are *not* worldless spectators who have the Earth as that which must simply answer questions of Reason's own determining. We are human, full of *humus.* Truly *human* understanding must have a certain *humility,* a certain aspect of not being the center of everything and the only voice worth heeding. It must orient to an ongoing conversation with the Earth, a conversation that *must be sustained if life is to go on.*

A final underlying theme that hasn't been said in just so many words is that Wittgenstein's notion of family resemblance can be taken rather literally. His image of a conversation between realms of discourse is a stunning image for the relation between adult and child, between teacher and student. It provides a way of recasting education as *neither* a matter of domination, mastery, and control, where children are nothing more than our silent "object" of inquiry and intervention, *nor* pedocentric compromise, where we remain silent in the face of some precocious, precious notion of "the child as learner." Education is a risky, tense *conversation* between the old and the young, between the old and the new. Given this, educational theorizing cannot be simply a matter of declaring an end to this conversation through objective re-presentations which render it univocal (i.e., turn it into the singular voice of the disinterested, methodical theorist). It is a matter, rather, of getting in on this conversation. Even though it is difficult, even though it is irresolvable, ongoing, fluid, and risk-laden, the conversation between teacher and student, between adult and child, is *not* something that needs fixing. We cannot "fix" the fact of natality. As educators, we cannot even imagine wanting to fix it. We cannot live in Descartes's dream, for in education, where we are constantly and essentially faced with difference, with renewal, with change, and with the full difficulty of conversing with children, such a dream cashes out as a nightmarish, inhuman vision aimed at rendering children silent and eliminating their vitality, their liveliness, their difference. The Cartesian desire to dispel ambiguity cashes out, in the end, as a desire to dispel the difference of children, to dispel our kinship with them. When Descartes's nightmare comes to haunt educational theory and practice, its hatred of ambiguity and difference in inquiry turns out to be a hatred of the generativity that issues from such ambiguity and difference. Descartes's nightmare becomes just this: a hatred of children.

Concluding Remarks: Educational Theorizing as Obedience and Thanksgiving

In his book *What is Called Thinking?* (1954, 1968), Martin Heidegger performs a telling reversal on the traditional philosophical question "What is called thinking?" Rather than entering the already crowded fray and proposing new and improved characteristics of thinking, Heidegger asks "What *calls* for thinking?" In this sudden reversal, the whole legacy of Descartes's nightmare is jeopardized. What Heidegger's work does is ask a simple question that has been lost in the wake of Descartes: What is it that thoughtful inquiry *heeds?* What is it that we are deeply responding to in education? What is it, in these times, that needs our response? What calls for thinking?

The difficulty with these questions—and this difficulty accounts for their seeming naivety—is that they are not concerned with the *methods* of thought or inquiry; it is not to method that these questions are obedient. They don't begin with a statement of position or purpose, a clarification of models, a description of procedures. They don't ask after what demands thought will make on things, but ask after what demands things make on thought. These questions thus highlight Descartes's nightmare—inquiry has become obedient only to itself, to its own methodical character to the extent that the question "What calls for thought?" appears simply naive. Amid the racket of research on every possible facet of the child's life, every possible facet of the curriculum, every possible facet of teaching practice, it is almost impossible to heed anything except this relentless, exhausting racket. There is no silence, no pause, no unprepared space in which something other than our own voices can be heard.

If we play for a moment with the etymology of "data," we find that it originally means "that which is given" or "that which is granted." Inquiry must open itself to that which is given or granted. It must be able to listen or to attend to that which comes to meet us, just as it comes to meet us. Inquiry need not prepare itself by arming itself with methods which demand univocity and clarity. Rather, it must do what it has always claimed to do—it must "gather" data. This metaphor should not be lost. What is given or granted is precious and delicate, and it must be gathered with all the love and care with which we gather the fruits of the earth, careful not to do violence, careful not to expect too much, prepared to wait, prepared—dare we admit it?—for the possibility that *nothing* will come forth (a possibility that teachers and parents live with all the time in living with children; such deep love, care, and risk only seem peculiar if one becomes cloistered in academia). This gathering cannot take its cue from our willingness to act without a moment's notice. It must take its cue from that which is given as to how the gathering must go. There is nothing we can do to *guarantee* that this gift will be given.

This, for me, is a holy metaphor for the act of teaching. We cannot *will* maturity in our children. We must wait, listen, and act exquisitely (i.e., act in response to what is called for, not just because we *can* act). To make our actions "exquisite," we must clear away the rattle and hum of unnecessary, uncalled-for action and noisy announcements, so that the need to act can begin to stand out. We must begin to believe again that silence may be our most articulate response. Silence must become possible again. In the midst of silence, a word, a gesture, a cry, can finally *mean* something, because we can finally hear, finally listen. The Latin roots of listening, attending, tell us even more than this: *ab audire*—to listen, to attend, is to be obedient. We cannot be obedient to that which comes to meet us by remaining blindly obedient to our methods for rendering what is given or granted into the sort of thing that is methodologically admissible. We must face children, seeing the actual living faces beneath the thin facade that our theories bestow.

Thoughtful inquiry is an attentive, and, one might say, appreciative response to that which is given, to this "gift." It is for this reason, perhaps, that Martin Heidegger (1954, 1968) rather playfully links up *Denken* and *Danken*—thinking, inquiry, becomes a form of thanks for what is given, a form of praise and song, a truly *human* voice:

> The things for which we owe thanks are not things we have from ourselves. They are given to us. We receive many gifts. But the highest and really most lasting gift given to us is always our essential nature, with which we are gifted in such a way that we are what we are only through it. That is why we owe thanks for this endowment, first and unceasingly (142).

Taking up this gift as something freely given means recognizing that it does not emanate from us; it does not fall within the purview of that which can be mastered and enslaved, bound in the frontiers of my Cartesian fear and aspiration. My own child is *other* than me, a precious gift of new life in the midst of my life. As an educator, and as a parent, I cannot dispense with the ambiguity and difficulty of such living in favor of the loneliness and despair of clarity and distinctness. Awakening from Descartes's nightmare is not a matter of simply developing a tolerance for ambiguity. It requires of us a love of ambiguity which is at once a love of the generativity of new life as a gift bestowed from the Earth. We can only hope that it is out of this love that educational theory and practice is born.

CHAPTER TWO

"The Fecundity of the Individual Case":
Considerations of the Pedagogic Heart of
Interpretive Work (1992)

Introduction

A former student-teacher phoned me in a panic late one August, excited that she had been offered a job in an Early Childhood Education classroom starting the next week and, of course, apprehensive about all that might entail. She phoned, I suspect, as much for reassurance as for advice. Eight weeks later, well into the school year, she phoned again and recounted the experience of going to her new school just days before the children were to arrive.

The principal was not available when she arrived, and she was instructed by the school secretary that her room was "down there, Room 10." She had walked down the hallway to what was to be "her room" and paused. The door was shut and she spoke of this shut door being "imposing," "as if something was going on in there already" that of which she was not yet part of, something to which she did not yet "belong." As she told it, she knew that when she opened that door, somehow, "everything would be different," things would be, in her words "turned around." She sensed that, once she "stepped in," she would be finally "crossing over" from student to teacher: "once I entered the room, I knew that would be *it.*"

We have all had similar experiences to this. In some sense, and to some degree, we all understand what she is talking about. Her tale is *familiar,* familial, something with which we *already* have deep, unvoiced kinship (Wittgenstein 1968, 36). In the face of this undeniable sense of kinship and understanding, what is the task of educational inquiry with respect to such an incident? How are we to do justice to this particular episode that happened to a particular teacher at a particular time and place, while at once respecting the undeniable kinship we experience in hearing this teacher's tale?

This paper explores how the interpretive disciplines understand and address the powerful "fecundity" (Gadamer 1989, 38) of such incidents. Understood interpretively, such incidents can have a generative and re-enlivening effect on the interweaving texts and textures of human life in which we are all embedded. Bringing out these living interweavings in their full, ambiguous, multivocal character is the task of interpretation. There is thus an intimate connection between *interpretation* (concerned as it is with the generativity of meaning that comes with the eruption of the new in the midst of the already familiar) and *pedagogy* (concerned as it is with the regeneration of understanding in the young who live here with us in the midst of an already familiar world) (Arendt 1969).

It is not simply that pedagogy can be one of the themes of interpretive inquiry. Rather, interpretation is pedagogic at its heart.

The first section of this paper is a playful consideration of unvoiced philosophical assumptions underlying those forms of educational inquiry which begin with methodical acts of severance in order to ensure "objectivity" in what they might have to say about such an incident. The next section shows how this incident could be read interpretively, bringing out the difference in the underlying assumptions of such an interpretive reading. The concluding section of this paper attempts to weave together more explicitly the threads that bind together interpretive research and pedagogy.

The "Isolated Incident" as the Substance of Inquiry

"A substance is that which requires nothing except itself in order to exist" (Descartes c.1640, 1955, 275) This is a long-standing definition, cited here from Descartes (seventeenth century) but winding its way back into the work of Thomas Aquinas (thirteenth century) and from there, back into Aristotelian metaphysics (third century B.C.E.). I cite it here because for much work in educational inquiry, the fundamental *given* (the root of the notion of "data" as "that which is given or granted") in inquiry is not that original, ambiguously alluring familiarity that first strikes us when we hear this teacher's tale. Rather, what is strictly *given* is the "isolated incident." The literal text produced by this particular teacher at this particular time in this particular situation—this is "that which requires nothing except itself in order to exist." This is the substance of (some forms of) inquiry.

Before we can begin such an inquiry, we must *make* this incident into something portioned off from anything else except itself. We must begin by systematic acts of severance aimed at retrieving the given ("the isolated incident") out of the amorphous web of interweaving meanings in which it was originally embedded. We must sever any interconnections that are already at work before the methods of our inquiry are enacted. We must (ideally, at least) put out of play any understanding of or connection to this instance that we may have as inquirers. We must suspend any spontaneous familiarity or sense of kinship that it evokes in us. We must also put out of play any interconnections we see or suspect between this instance and any other meanings or tales or stories or narratives.

These two acts of severance—this instance from us and our lived familiarity with it, and this instance from other instances—will allow it to become a self-identical substance, something that stands "without us" and without reference to any other incident. Thus severed, it no longer signifies or signals anything beyond itself. It becomes, as we know thus far, "an isolated incident," just itself and nothing more. These systematic severances have acted on the *assumption* (implicit in empiricism) that all that is *given* is the empirical instance. Therefore, any interconnections or evocations have been imposed upon it and these impositions must be put out of play before we can retrieve the integral instance itself. Our isolation of the instance, then, is done against the backdrop of the belief that it is "in fact" isolated. In this way, our methodical severances are not understood as violations of already-existing, real, and vital interconnections. Rather, these severances involve systematically *reversing* those violating interconnections that have despoiled the actually isolated incident. We have retrieved the integrity of the instance by retrieving the isolated, individual (i.e., not further divisible) case.

This is a fascinating process to which we subject both the instance and ourselves. It is akin to a sort of purification ritual (Bordo 1987, 78–82) that both we and the instance must undergo. Regarding the instance itself, ambiguous linkages and tell-tale signs and marks of potentially violating interconnectedness are systematically eliminated, producing of a sort of virginal, untouched instance. And regarding ourselves, we can no longer approach this instance with the moist and fleshy familiarity with which we began. We must now simply "behold" it with what Alfred North Whitehead named the "celibacy of the intellect" (cited in Fox 1983, 23). We must remain strictly within the parameters of the methods of severance we have enacted, for any other interconnection would despoil or defile the instance we have so carefully and methodically isolated and purified. Our connection to this instance thus becomes gutted. We understand it "from the neck up," uprooted from the dark and original familiarities and kinships which have been put out of play. And, correlatively, the instance itself loses its ambiguous allure and is rendered fully *present.* Along with the assumption that all that is given is the isolated instance, we find a correlative assumption: the *given* is equatable with *the clear and distinct.* Any signs of ambiguity in what is given (in "the data") indicate that we have not yet rid the given of its impurities or not yet controlled for the possible interpenetrations of dependent and independent variables. The isolated instance, if properly isolated, is what it is and therefore can contain no ambiguity. Ambiguity or any other sign of a lack of clarity and distinctness is understood to be nothing more than a problem that needs to be fixed through further purifications and severances. An ambiguity in the data is thus simply the occasion to subdivide the problem and conduct a further study. The given, therefore, is univocal, clear, and distinct. Any entrails of meaning that might have wandered from it off into dark corners or that may have dug deep into our lives and drawn us into unanticipated, illicit interplays have been cut off.

A more direct way of putting this process is that, through these severances of the original familiarity in which we were immersed and which drew us in in the first place, we render this instance into an *object* and, correlatively, render ourselves into a "knowing subject" which has this object, not as something to which we belong and have a kinship or relation, but as something standing over against us. The instance-as-object now no longer *fits* into a complex fabric of interrelations in which I belong with it, but rather "stands out," isolated from what surrounds it. It becomes "obtrusive, importunate, and demanding of our attention" (Weinsheimer 1987, 5).

From this original severance thus begins a long series of correlative movements between this instance and myself as inquirer. "Subject and object precipitate out simultaneously. Yet even while separate, they remain interdependent, because the breakdown in the world [i.e., the tearing of the instance out of the fabric of familiarity in which it originally lived] corresponds to a breakdown in understanding" (Weinsheimer 1987, 5). Once divested of the original, intimate knowing, I can no longer claim to understand this now severed object. In this way, "both subject and object are derivative and secondary, in that both precipitate out of the more primordial unity of being at home in the world" (5), a "being at home" bespoken by the fact that I somehow "already understood" what this teacher said *before* the specific work of rendering it an object of research even began. This precipitate subject and object "are [both] determined negatively: the knowing subject [now severed from our original senses of familiarity] no longer understands and the object [now severed from its living context] no longer fits" (5).

These fundamental acts of severance and the convoluted sequence of correlative "purification transformations" in both the object of inquiry and the inquirer give inquiry a peculiar rootlessness. Once we become severed from the abiding senses of kinship and familiarity and embodied allure that this instance evokes (once it becomes an "object" and we become a "knowing subject"), we are left with clear, univocal, given surfaces *both* regarding the instance and regarding ourselves. *It* is transformed into what objectively presents itself to us (i.e., univocal "key terms," or coded words, that can be accurately mapped and charted) and *we* are transformed into deployable methods that themselves have a clear and univocal character (Weinsheimer 1987, 6). Once these instances of our lives become uprooted from their fitting place in the world and once we become uprooted from our familiarity with the world, inquiry into such (now "objective") instances becomes enamored of *frequency* and *reoccurrence.*

The only significances we can glean from these rootless surface readings of the incidents of our lives are from quantities and enumerable surface repetitions. When, for example, we hear a beginning teacher talk about the anxieties of opening the classroom door for the first time and entering in, speaking and writing of the resonant *meaning* of such an event is foregone in favor of an inquiry into whether a significant *number* of "respondents" will cite the same experiences, use the same words and concepts, speak in the same terms in their reports. Because we have actively and intentionally reduced this instance to an isolated incident, it becomes essential to collect more and more incidents in order to raise this first incident out of its isolation. Because we have actively and intentionally restricted ourselves to that knowledge produced methodically, it becomes illegitimate to engage these instances in ways other than simply collecting them. This first instance becomes significant (that is to say, it points to something beyond itself) only insofar as it can now be shown to *reoccur* in a (mathematically) significant number of other equally actively isolated incidents. Significance thus becomes intimately linked with *frequency.* More pointedly put, *significance becomes mathematized.* This instance links up with others only under the watchful eye of this most celibate of disciplines.

The interest of such a mathematization of significance is not to better *understand* this instance and its meaning as a feature of human life, but to be better able to control, predict, and manipulate its future reoccurrences (Habermas 1972). Above, we mentioned that, following upon the methodical severances of our familiarity with the world, there is a correlative negative determination of both object (which now no longer fits) and subject (which now no longer understands):

> The cognitive remedies for these twin defects are likewise correlative. The object is disassembled, the rules of its functioning are ascertained, and then it is reconstructed according to those rules; so, also, knowledge is analysed, its rules are determined, and finally it is redeployed as method. The purpose of both remedies is to prevent unanticipated future breakdowns by means of breaking down even further the flawed entity and then synthesizing it artificially. Thus Gadamer speaks of "the ideal of knowledge familiar from natural science, whereby we understand a process only when we can bring it about artificially" (1989, 336) (Weinsheimer 1987, 6).

Once these "cognitive remedies" are enacted, we can (within mathematically prescribed limits) predict the reoccurrence of such incidents and therefore we no longer be "taken aback" by such reoccurrence. Such incidents will not allure us again and catch us off guard, with all the disorienting and disturbing consequences that such allure can have. These remedies (recall, *produced* of the original precipitation of "subject" and "object") prevent the possibility of understanding being *provoked by something* unwittingly and without methodical anticipation. Thus, "objectification" protects us from dangerous unanticipated turns that the world may take (this is precisely the strength of such work). It rules out of its considerations unanticipated ("uncontrolled-for") interchanges with the world.

Of course, the methodical attainment of such objectivity does not altogether prevent playful, risk-laden, unanticipated interchanges. They will still *occur*. However, their occurrence is divested of any claim of or access to *truth*. Truth and method become identified. It is precisely this identification that the interpretive disciplines work against. Certainly the methods of quantitative research can help us better understand this incident and their assertions can make a claim to truth. The interpretive disciplines suggest, however, that there is a "truth" to be had, an understanding to be reached, in the provocative, unmethodical incidents of our lives, a truth which is despoiled and thus left out of consideration by the methodical severances requisite of empirical work.

An Interpretive Reading of the Instances of our Lives

The term "initiation" in the most general sense denotes a body of rites and oral teachings whose purpose is to produce a radical modification . . . of the person to be initiated. Initiation is equivalent to an ontological mutation of the existential condition. The novice emerges from his ordeal a totally different being: he has become *another* (Eliade 1975, 112).

"I knew when I walked through that door, *I* would be the teacher. Everything would be different." Perhaps this teacher's words can be read as a retelling of the grand narratives of initiation and transformation, "insiders" and "outsiders," thresholds and boundaries, of being turned around in those moments when everything becomes different, of risking self-understanding and self-definition by moving into a new sphere, of repetition and renewal, of the turns and interplays of responsibility and irresponsibility, of the turns from childhood to adulthood.

Interpretive research begins with a different sense of *the given*. Rather than beginning with an ideal of clarity, distinctness, and methodological controllability and then rendering the given into the image of this ideal, it begins in the place where we actually start in being granted or given this incident in the first place. It begins (and *remains*) with the evocative, living familiarity that this tale evokes. The task of interpretation is to bring out this evocative given in all its tangled ambiguity, to follow its evocations and the entrails of sense and significance that are wound up with it. Interpretive research, too, suggests that these striking incidents make a claim on us and open up and reveal something to us about our lives together. In this sense, our unanticipated, unmethodical being in the world can, quite literally *in* certain instances, make a claim to truth.

When this teacher phoned me, her words evoked in me a sense of something *already familiar* that I did not fully understand, but somehow undeniably "knew." I felt suddenly implicated by her words, as if she spoke about something in which I was somehow already involved and which I somehow already understood but had forgotten or not explicitly noticed. Interpretive inquiry thus begins by being "struck" by something, being "taken" with it—in this particular case, the unanticipated eruption of long-familiar threads of significance and meaning in the midst of a wholly new situation. "Understanding begins . . . when something addresses us" (Gadamer 1989, 299). This striking incident *called for* (Heidegger 1954, 1968) understanding. For all its incidentalness, it aroused and generated a new and fresh understanding of something already understood. It opened up something that seemed "over and done with."

It is at this juncture that the true fecundity of the instance comes into play.

This teacher's story is not an isolated instance to which the concept of "initiation" is to be applied, as if "initiation" were already understood, already fixed and closed and definitively defined, and this instance were simply a replica or a copy of it. Rather, what this teacher's story speaks of *is* initiation—it *belongs to* initiation and therefore adds itself to what initiation can now be understood to be. But saying that this instance *is* initiation requires understanding "is" in the manner of *analogia entis:* in the manner of "analogical being." Its *being* initiation does not mean that it is *identical* in all respects to some pregiven and preunderstood fixed set of concepts (this would make the instance superfluous to this already-established meaning). But neither is this instance simply "nothing except itself," simply *different* than initiation. Rather, the instance is, so to speak, the generative offspring or "kin" of initiation. It bears a "family resemblance," (Wittgenstein 1968) to initiation, interweaving with it in ambiguous ways that are not mathematizable into univocal terms that could be simply counted and recounted. For with this teacher's tale, it is not perfectly clear whether we have an unambiguous reoccurrence of some phenomenon, for this tale is not *identical* to any other instance of initiation (but neither is it simply *different*).

What we have, rather, is something vaguely familiar, vaguely recognizable, something that bears a "family resemblance" that warrants further investigation.

Thus, the relation between the instance and that to which it seems to bear a "family resemblance" is always in a type of suspense. Interpretive inquiry does not wish to literally and univocally say what this instance *is*. Rather, it wishes to playfully explore what understandings this instance makes *possible*. There is not a question, then, of whether this instance "really is" an instance of initiation, but whether it is possible to understand it this way and whether understanding it this way is "fruitful." It justifies this approach by harkening back to the fact that it does not take up this instance as an "object" with certain given characteristics. It takes up, rather, as something which evokes and opens up an already-familiar way of belonging in the world, a possible way of being (i.e., "being an initiate"). This instance must be taken up as a "text" which must be read and reread for the possibilities of understanding that it evokes. Interpretation involves "making the object and all its possibilities fluid." (Gadamer 1989, 367) That is to say, interpretation "make[s] the novel [this particular incident] seem familiar by relating it to prior knowledge, [and] make[s] the familiar [what we have already understood " initiation" to mean] seem strange by viewing it from a new perspective" (Gick & Holyoak, 1983, 1–2). Thus, interpretive work doesn't simply read the instance into a pregiven, closed, and already understood "past," but, with the help of the instance, makes what has been said of initiation in the past *readable* again by reopening it to new, generative instances. To the extent that interpretation makes things readable, it is intimately linked up with a sense of literacy.

This particular instance, then, can be *understood* as bearing forward the phenomenon of initiation, reinvigorating it and thus transforming it, making it fruitful, making it a forebearer. Initiation thus *needs* the instance to become and remain generative. Put the other way around, without living instances, initiation would no longer be a living feature of our lives; it would no longer be something that concerns us, that provokes us, that entices us. Initiation would no longer be an ongoing, vibrant narrative or story of which our lives and our experiences are an intimate part and to which we belong. It would simply be a lifeless concept or the name of some object which "stands apart" from the life we live, couched in some textbook, an object of indifference.

It is in this sense that the instance is fecund: it *keeps the story (of initiation) going,* a keeping going which *adds to* the story and which thereby changes what we will come to understand the already past chapters to have meant. What we have with interpretation is a process akin to having children. The birth of my son transformed me into being a father, and my father into being a grandfather. Paradoxically then, my son regenerated what I have come to understand the course of my life to have *already been.* He constitutes not simply the addition of one new, isolated element in a chain of events. He constitutes the necessity to rethink the whole chain and each event in it. Thus we can legitimately speak of the "fecundity of the individual case" insofar as it is allowed to wind its regenerative tendrils out into the "old growth" from which it has erupted—insofar, that is, as we do not begin our work by severing precisely these regenerative tendrils of sense.

We end up, here, with one of the most telling features of interpretive work. Initiation is not a *given* whose features can be simply listed and to which instances can be simply compared. Rather, the relation between initiation and the instance is an interpretive one. The new instance transforms what initiation is, and initiation helps articulate what the instance means. The instance is thus irreplaceable in its particularity, because that very particularity can have a generative, transformative effect that cannot be duplicated. It is this resistance of the particular to simple, powerless subsumption that helps interpretive inquiry from simply being a reiteration of conservative, traditional understandings. Those shared and contested understandings in which we live are *called to account* by this instance, made to "speak," change, accommodate, and, so to speak, "learn" through this encounter.

If an instance is simply duplicated in all respects over multiple cases, the duplicates add nothing new to our understanding of what initiation *is*. They will simply confirm its reoccurrence. More simply put, a quantitative study may provide us with irrefutable assurance that the phenomenon of "initiation" is widespread among beginning teachers (a valuable piece of information in and of itself). But it can accomplish this without adding *at all* to our understanding of what initiation *is* and what it *means* as a feature of human life. It is *this*—adding to our understanding of our lives—that is of interest to interpretive inquiry.

This "adding to the understanding of our lives" is not a matter of establishing once and for all what certain objective features of human experience are and are not. We cannot fully know once and for all what "initiation" is because, so to speak, it *is not* yet. As something which forms a living part of our experience, we don't fully know what it *is* because we don't yet know what will become of it. And we don't know this *because it is still coming.* To the extent that we do not know what is to be made of initiation in the future—how it might appear and how those appearances might transform our understanding of what it means to *be* an initiate—our interpretive relation to this particular instance cannot be oriented towards having some "last word" about it *as if* it were an "object" that is simply present, that simply stands there before us to be univocally named. "It would be a poor hermeneuticist who thought he could have, or had to have, the last word" (Gadamer 1989, 579). A "good" interpretation, then, is not definitive and final, but is one that keeps open the possibility and the responsibility of *returning,* for *the very next instance* might demand of us that we understand anew. It is for this reason that the language of interpretive inquiry can be so unfailingly annoying, for it purposely struggles against the tendency of language towards literalism and univocal declarations regarding what is and is not the case. Its language tends, therefore, to be more "playful" and seemingly less serious than other forms of inquiry. However, there are serious consequences at issue. Failing to keep open the possibility of returning to understand anew is at once demanding of initiation that it no longer be open to the possibility of fecund new instances. It renders it into a fixed object that needs nothing except itself in order to exist. It is equivalent to believing that the new, the young, have nothing new to add. It is equivalent to believing that the human story can go on without renewal and regeneration.

How do I *know* that this reading I have given this instance is reliable? Well, I *don't know all by myself* or in advance of the reading itself. I cannot separate out in advance which features of my reading reveal nothing more than idiosyncrasies of my individual experiences and which features reveal something more—not the teacher's text "in itself," but the binding arcs of meaning in which I, that teacher, and the text belong together. I cannot separate out in advance and by some pregiven method whether the work of Mircea Eliade I read as an undergraduate student or Johan Huizinga's texts on play, or my own life experiences, will end up having any fruitful bearing on my reading of this instance. "This separation must take place in the process of understanding itself" (Gadamer 1989, 296). And this separation will take place only if I let my preunderstandings fully *engage* this text; I must let them be brought fully into play and therefore risk that they might be changed in confronting what this teacher's text has to say (Gadamer 1989, 299).

Put more sharply, for interpretation to engage, the text and I must be allowed to "play." And *in* such play, an unavoidable paradox of interpretive work comes to light. The fact that I happened to have read and remembered Mircea Eliade's work on initiation, the fact that I happened to have been called by this teacher and to have been struck by what she said—all of these "happenstances" made possible the interpretation that will then ensue (Weinsheimer 1987, 7–8). The interpretation is thus unavoidably linked to *me*. It is not something produced by a method that anyone could wield. However—and here is the paradox—what the interpretation is henceforth *about* is not me and my past experiences, but that *of which* I have had certain experiences: initiation. Even though interpretive work is not possible without a living connection to its topic, it is *the topic,* not the *fact* of a living connection, that is the center of interpretive work. The same can be said of the reader of an interpretive study. If the reader has no living connection to or experience of something like the phenomenon of initiation, the study will be meaningless, for it will not address something to which the reader bears any "family resemblance."

Producing a "reliable" interpretive reading of this instance requires living with this instance for a period of time in order to learn its ways:

turning it over and over, telling and retelling it, finding traces of it over and over again in what you read, seeing the nod of heads and faint smiles when it is used as an example in a class, scouring the references colleagues suggest, searching my own lived-experience for analogues of experience, asking friends if they have experienced anything like this before, testing and retesting different ways of speaking and writing about it to see if these different ways help engage and address possible readers of the work to follow. It takes time to dwell with such an incident and allow the slow emergence of the rich contexts of familiarity in which it fits. I can learn the ways of this instance only by taking the time to experience where it "goes," and thereby seeing to what territories and terrains it belongs. This instance is thus not static but rather "leads" somewhere. Time is needed to pare it down to what readings might be fitting of its ways, but this time, in an important sense, *belongs to the instance itself.* In spite of my deadlines and desires, very often fitting insight "takes its own sweet time." Only over this unmethodical course of time does the full fecundity of the individual case come forward. And, some might say unfortunately, "there is no art or technique of happening onto things. There is no method of stumbling" (Weinsheimer 1987, 7). This incident, which gave rise to so much, *just happened.*

For a reliable reading to occur, then, it would never be enough to simply say what I think it means and leave it at that. But neither is it enough to simply turn it back to the "respondent" and ask whether she *intended* to mean something about "initiation" or "responsibility" or "becoming an adult" and the like, as if calling out to the author might save us the task of interpreting the text. The author's (respondent's) reading of her own story is not the lynchpin of hermeneutic work (as it might be for some forms of "teacher narrative" now gaining ascendancy in educational inquiry). Rather, "we are moving in a dimension of meaning that is intelligible in itself and as such offers no reason for going back to the subjectivity of the author" (Gadamer 1989, 292). The living, generative meaning(s) of the text are at the center, and the game of interpretation is afoot for us all in the face of this text.

> Language . . . is by itself the game of interpretation that we all are engaged in every day. In this game nobody is above and before all the others; everybody is at the center, [everybody] is "it" in this game. Thus it is always his turn to be

interpreting. This process of interpretation takes place whenever we "understand."
(Gadamer 1977, 32)

This does not mean that the connection to the author is severed. It means, rather, that, in the face of this text, the author is *one of us* and not in some elite, "authoritative" position. Certainly, the author is in an elite position regarding the *experiences* she underwent, but, once erupted into a text, she is not in such a position in regard to what those experiences might *mean*. In this particular case, in fact, this teacher was relieved to discover that her experiences were not just hers, that what she was going through *meant* something to those with whom she spoke. The expression of her experience into a text thus relieved her of the burden of isolation. She discovered that her experience linked up with long-standing characteristics of human experience and articulation. She discovered that this experience had a character and vitality over and above the fact that she had undergone this experience and the fact that it had been powerful for her. This discovery, as mentioned above, puts a peculiar spin on the notion of "literacy." Decoding, counting, and recounting the surface signs (of text, of experience) is not enough, rather, as we unearth the signs of life crackling underneath the surfaces, "we . . . become more literate [and] we may become less literal, [less] stuck in the case without a vision of its soul" (Hillman 1983, 28). Simply "telling my story" can unintentionally breed a type of literalism/illiteracy by disallowing "a vision of its soul."

Such a "vision" would help liberate my story from being *just* mine (which bears a frightening resemblance to the severances and isolations requisite of objectivity). This is why language plays such a predominant role in interpretive work, for, by its very nature, it serves to raise up the instances of our lives out of the burden of their specificity (Gadamer 1989; Smith 1991). It allows us to escape "the compulsive fascination with one's own case history" (Hillman 1987, 7). It makes it possible to see what one is going through as intimately wound up in human life as a whole, a generative "process that is continually internalizing and externalizing, gaining insight and losing it, deliteralizing and reliteralizing" (Hillman 1983, 27). It thus allows us to read our individual lives as fully participant in the shared and contested, generative work of humanity as a whole.

Thus, in interpretive work, the author's reading is but one voice among many, *perhaps* an especially unmindful and unattentive one, *perhaps* the one best suited to read this text well. Understanding this teacher's text, then, is not a matter of unearthing her experiences, but of "clarifying this miracle of understanding, which is not a mysterious communion of souls, but sharing in a common meaning" (Gadamer 1989, 292). "What emerges," in opening up a conversation with this instance "is neither mine nor yours" (331) but is that "in which" we dwell together—the contours of that original familiarity and kinship that made this instance so telling in the first place. "Understanding is the expression of the affinity of the one who understands to the one whom he understands and to that which he understands" (Gadamer 1983, 48). None of us necessarily knows all by ourselves the full contours of the story each of us is living out. This is why dialogue and conversation figure so predominantly in interpretive work, as contrasted with the "monologue" of scientific discourse (Habermas 1972), suitable as such a monologue is to the univocal character of "isolated incidents" and the correlative univocity of the methods deployed by a "knowing subject."

One problem, of course, is knowing when to stop in the spinning out of implications of meaning. There are widespread possibilities embedded in this incident and there is no surefire method for guaranteeing that you haven't gone too far and stretched the incident out of all proportion. For example, when we begin to picture this incident as a retelling of the tale of initiation and then couple it with an innocent comment that I have heard from several principals regarding student-teachers and beginning teachers, sparks begin to fly: "I like having student-teachers in my school because the profession constantly needs new blood." The connection of "blood" with the rite of the initiation of the new ones into the profession (which literally means "those who take to vows," yet another feature of initiation rituals) becomes even more telling when we recall that rites of passage and initiation tend to take place in the spring—the time of Easter (itself a sacrificial blood ritual involving the opening of barriers and allowing the ones outside to come in), of graduation, as well as the time when interviews are often done for school boards seeking "new blood." At the tail end of this sequence, we may have gotten rather "carried away," but the implications are not meaningless. In spite of the fact that they can easily become too "wild," they are not altogether "unfitting." The "analogical kinships" of meaning (Jardine, 1990) still seem to pertain. The "family resemblances" persist.

The problem of interpretive research, then, is one of withholding the interpretive impulse and developing a sense of proportion in discourse. Again, this is not a method that can be handed over (it is almost impossible to answer a question like "How do you do hermeneutics?"), but is a practice. It is a practice in a strong sense precisely because the incident under consideration and the concrete context of speaking about it (with this beginning teacher in the midst of her anxiety, in casual conversation with a friend, as a topic in a class, as a subject for an academic paper, etc.) will have something to say about what a "good" sense of proportion might be in this case or that. One cannot say, therefore, in general and ahead of time, what the practice of interpretation is like, as if it were a set of rules that needed to be simply *applied* to an incident independently of the contribution that incident might have regarding what needs to be said (Smith 1991). This point, again, bespeaks "the fecundity of the individual case" in the pursuit of understanding.

Concluding Remarks: On Interpretation, Pedagogy, and Hermes as Trickster and Thief

Hermes is cunning, and occasionally violent: a trickster, a robber. So it is not surprising that he is also the patron of interpreters (Kermode 1979, 1).

When Hermes is at work . . . one feels that one's story has been stolen and turned into something else. The [person] tells his tale, and suddenly its plot has been transformed. He resists, as one would try to stop a thief . . . this is not what I meant at all, not at all. But too late. Hermes has caught the tale, turned its feet around, made black into white, given it wings. And the tale is gone from the upperworld historical nexus in which it had begun and been subverted into an underground meaning (Hillman 1983, 31).

There is one further aspect of Hermes that may be worth noting, namely his impudence. He once played a trick on the most venerated Greek deity, Apollo, inciting him to great rage. Modern students of hermeneutics should be mindful that their interpretations could lead them into trouble with the authorities (Smith 1991).

It is admittedly rather frightening and disorienting to discover that the incidental story we might tell can have implications of sense that we did not anticipate and cannot fully control. We all know, and have all suffered in our own ways, how this has often meant that others have spoken in our stead and "for our own good."

But to say that Hermes is a trickster and a thief is not to say that the inquirer is Hermes and the teacher I spoke with is the sole victim of the theft of meaning and the subsequent transformations of understanding that ensue. Rather, the playful tricks and turns happened to me as much as to her. It is not as if I could, in the inquiry I pursued, say anything I wanted and do anything I wished. I, too, was "subject" to Hermes's seeming whims, having things collapse without warning, gaining insights at the worst of times and losing them before I could catch them, muttering quite often while writing or rereading what I thought was so clear "this is not what I meant at all, not at all. But too late . . ."

Pursuing interpretive inquiry is a potentially painful process, because it is not produced of a method which (ideally) will keep everything under control by severing all the tendrils of sense that can pull you in so many different, often incompatible ways. There is a risk involved in such work, a risk of "self-loss" (Gadamer 1977, 51) and the recovery of a sense of self that is different than the one with which we begin such inquiries.

There is a straightforward sense in which interpretive work is pedagogic: it is concerned with the regeneration of meaning and is therefore disruptive of fossilized sedimentations of sense, desiring to open them up and allow "the new" to erupt and thus allowing the old and already established and familiar to regenerate and renew itself. It is oriented to "keeping the conversation going" (Smith 1988b) between the new/young and the established/old about the texts and textures of human life. It moves against the *stasis* inherent in objectivism and literalism.

But there is a different sense in which interpretive inquiry is pedagogic. The process of interpretation is not the simple accumulation of new objective information. It is, rather, the transformation of self-understanding. Living with this instance and following its ways and engaging my own life and the lives of others in an attempt to understand it has changed who I am and what I understand myself to be. New possibilities of self-understanding have opened up; old ones have been renewed and transformed and rejected. What I understand myself, my work and the lives of my students to be have changed, for better or worse. And, of course, all these understandings cannot now be trumpeted as final. They will have to work themselves out over the course of my life and the lives of those I engage.

Moreover, writing of this incident is not a matter of passing on information to a reader, but of evoking or educing a different self-understanding in the reader. The goal of interpretive work is not to pass on objective information to readers, but to evoke in readers a new way of understanding themselves and the lives they are living. Following the entrails of sense that this incident regenerated means, in however small a way, understanding who we are differently, more deeply, more richly.

Unlike some work in educational inquiry which begins with a "knowing subject" that is fully in possession of itself ("itself" being defined as the methods it can deploy), interpretive work inevitably begins with a living subject in a living dialogue with the life that surrounds us. "To reach an understanding . . . is not merely a matter of putting oneself forward and successfully asserting one's own point of view, but being transformed into a communion in which we do not remain what we were" (Gadamer 1989, 379). In such a case, interpretive work is profoundly pedagogic, for:

> in the last analysis, *all* understanding is self-understanding, but not in the sense of a preliminary self-possession or of one finally and definitively achieved. For .
> . . self-understanding only realizes itself in the understanding of a subject matter and does not have the character of a free self-realization.
>
>   The self that we are does not possess itself; one could say that it "happens" (Gadamer 1977, 55).

Afterword

One more playful turn. "Understanding is an adventure and, like any other adventure, it is dangerous" (Gadamer 1983, 109–10). Involvement in interpretive inquiry runs the risk of getting quite lost in the flurries of sense that make up our lives. It faces, too, the dangerous insight that, so to speak, "getting somewhere" in understanding one's life is never finished—understanding "always must be renewed in the effort of our living" (110–11) and this need for renewal is not an *accident* that we can fix, but a situation that we must learn to live with well.

In the end, the notion of "initiation" is of especial interest to hermeneutics. Hermes, as mentioned above, was a trickster and a thief. He was also a messenger, a "go-between." Initiation has to do with the rites of passage and with transformations in how we understand ourselves, and this is precisely the interest of hermeneutics. Hermes is identified with borders, with boundaries and with keeping open the gates between one realm and another: "to hear the messages in whatever is said. This is the hermeneutic ear that listens-through, a consciousness of the borders, as Hermes was worshipped at borders. Every wall and every weave presents its opening. Everything is porous" (Hillman 1987, 156).

CHAPTER THREE

On the Humility of
Mathematical Language (1990)

Introduction

Thinking is not a means to gain knowledge. Thinking cuts furrows in the soil of Being. About 1875, Nietzsche once wrote (Grossoktav, WW XI, 20): "Our thinking should have a vigorous fragrance, like a wheatfield on a summer's night." How many of us today still have the sense for that fragrance (Heidegger 1957, 1971a, 70)?

How peculiar it seems to consider this passage as offering images of the thinking and language of mathematics. Mathematical language appears unearthly, virginal, born of what Alfred North Whitehead called the "celibacy of the intellect" (Fox 1983, 24). It is considered a serious and exact science, a strict discipline, and such images of seriousness, exactness, and strictness often inform how it is taught and how it is understood.

In the face of such persistent images, mathematics has become simply meaningless for some teachers and some children, producing little more than anxiety, apprehension, and the unvoiced belief that mathematics is a matter for someone else, for some "expert" who has abilities and understanding which are "beyond me." It has become inhuman, lacking *humus,* lacking any sense of direct presence in or relevance to our lives as they are actually lived. It seems that it still is, as it was for the ancient Greeks, a divine science that knows no humility, no place in the moist darkness of the Earth.

In this article, I want to trace a line of thought aimed at reembodying mathematical discourse into the discourse of everyday life. The claim of this paper is that the Earth and our everyday language and experience resonate with an ambiguous presence of *ma themata* —with the themes, the lessons, the interconnections of which the mathematics curriculum is but a ghostly idealization that has forgotten its origins, its *humus,* its humility.

Such was the intent of Edmund Husserl's phenomenology—to describe the deep embeddedness of the "exact" sciences in the life-world, in life as it is actually lived. He maintained that we cannot understand the discourse of the sciences by beginning with the "surreptitious substitution of [a] mathematically substructured world of idealities for the only real world . . . our every life-world" (Husserl 1932–34, 1970a, 48–49). If we begin with such a substitution, the resonances of mathematical discourse in everyday life end up being formulated as simply a blurring of what is in fact clear, a concretizing of what is in fact abstract, a making profane of the sacred— humiliation.

Phenomenologically, the reverse is the case. The idealizations of mathematical discourse appear in the midst of the world of everyday life, and they are not despoiled by such appearance but enlivened by being connected back to their living sources—"these are *human* formations, essentially related to *human* actualities and potentialities, and thus belong to this concrete unity of the life-world" (Husserl 1932–34, 1970a, 170). Mathematical discourse resonates deeply with our humanity, with our language, with our Earth. It is, in vital and important ways, child's play, with all the joy, difficulty, immediacy, and humility that such play invites. And again, this does not despoil its idealized exactness. Rather, it makes such exactness a *real achievement* that erupts out of life as it is actually lived, rather than seeing such exactness as graciously bestowed "from above." Mathematics is not something we have to look up to. It is right in front of us, at our fingertips.

This article begins with a reflection on an undergraduate Early Childhood Education curriculum class, where the topics of higher and lower numbers and symmetry were discussed. This is followed by a brief examination of the notion of analogical discourse as a way of understanding the decidedly human character of mathematical language. Finally, consideration is given to the way in which analogical language involves a recovery of the familiar, the familial, those "family resemblances" or "kinships" (Wittgenstein 1968, 33). that bind us to the Earth and to each other, matters that cannot be exhausted or clarified once and for all by pontifical curricular declarations. Analogical language opens us up to the generativity that lies at the heart of education itself and therefore lies at the heart of the mathematics curriculum as a *curriculum vitae.*

Giving and Drawing Boundaries:
A Class in Early Childhood Curriculum

To undergo an experience with language . . . means to let ourselves be properly concerned by the claim of language by entering into and submitting to it. If it is true that [we] find the proper abode of [our] existence in language—whether [we are] aware of it or not—then an experience we undergo with language will touch the innermost nexus of our existence (Heidegger 1957, 1971a, 57).

If we may talk of playing games at all, it is not we who play with words, but the nature of language plays with us, long since and always. For language plays with our speech—it likes to let our speech drift away in the more obvious meanings of words. It is as though [we] had to make an effort to live properly in language. It is as though such a dwelling were especially prone to succumb to the danger of commonness. Floundering in commonness is part of the dangerous game in which, by the nature of language, we are the stakes (Heidegger 1954, 1968, 18–19).

In a recent undergraduate class in Early Childhood Curriculum, I asked my students the following question: "In precisely what sense is 198 a *higher* number than 56?" The initial reaction to this question was silence, followed by scattered bewilderment and confusion. Although my students are becoming accustomed to this sort of question, the precise intent in asking it was not clear. Some students took the question as an indirect form of accusation—198 isn't "really" higher than 56, so the fact that they may have been using this language is an error to be corrected. Others simply struggled to make explicit what would be meant by "higher." They found themselves caught up in a swirl of interweaving and interconnecting meanings which seemed to resist being "straightened" out in any definitive manner. One student slipped into the language, common to young children, of numbers being "big" and "little." Far from remedying our situation, it simply multiplied the problem, so to speak.

The question then arose: If we don't know precisely what we mean when we use such language, how is it that we can feel confident when we attempt to teach such aspects of mathematics to young children? Implicit here is the equation of the ability to teach something with knowing what it is that you are teaching. This equation is one which I tend to encourage in my students. However, there is a deeper supposition here that must be addressed.

There is implicit here the equation of "knowing what it is you are teaching" with being able to be precise, to be exact and fully explicit, to provide foreclosed, literal definitions, and the like. One of the points I hoped to educe with my demonstration was that we *do* know what it means to say that 198 is "higher" than 56, but that this knowing is not definitional, literal, univocal, or clear. It interweaves in unanticipated ways, with the young child building a higher and higher tower of wooden blocks, with the fact that we can speak meaningfully of "counting *up* to ten" or with the fact that growing older means growing "up," and growing up means becoming taller, and that the "higher" one's chronological age, the "bigger" one is, and that, for children, importance bears a resemblance to height and age, and so on. The initial difficulty with such interweavings is precisely this "and so on." Although reflecting on our language can bring forth unanticipated, playful interweavings of experience, it is never quite clear, in following such interweavings, if one has gone too far. It is never quite clear just what the parameters are. After all, is it too much to say that the progression of higher and higher numbers orients to infinity (i.e., to God, the most High), and that numbers that fall below the "ground" (below where we stand, below "[ground] zero") have a dark and negative character? Or is it too much to say that when counting higher and higher quantities, we must keep track of them by consistently bringing them back to Earth, back to base, so that we use "base ten" as a way of preventing the pile from spiralling upward out of sight, a way of keeping them at our fingertips (our digits?) that we therefore organize higher and higher quantities into groups we can manage or handle, into "handfuls"?

Clearly these examples "go too far," but they are not altogether meaningless. Something is at play which belies the literalism and foreclosure of the mathematics curriculum guide, and it is this "something" that was at work in our class. Although there is indigenous to the discipline of mathematics a form of literalism and exactness of speech (which curriculum guides in elementary mathematics seem to emulate), it is not this exactness that makes it possible for one mathematician to understand what another is talking about. The telling point in this demonstration is that even in the use of terms like "higher" and "lower," we *already* understood what each other was talking about without aspiring to literalism and exactness. Understanding each other, then, was not founded upon literalism or exactness or definition, but was found to be already at work. Far from simply "playing with words," we had come upon deep ways in which our mutual understanding was *already* informed and made possible by interweaving meanings and implications of the language we were using as a matter of course.

As Ludwig Wittgenstein noted, we can *draw* a boundary for these concepts, "but I can also use [them] so that the extension of the concept is *not* closed by a frontier" (Wittgenstein 1968, 33). Regarding the desire we may have to define the term "higher," to draw a boundary or frontier, Wittgenstein rather playfully says "that never troubled you before when you used the word" (36). For example, we have all said, without hesitation and without trouble, that piles of pennies are the same size/height, implicitly meaning by this "the same quantity" and knowing full well that this is not a matter of the *equation* or *identity* of "height" and "numerical quantity" in all instances. The attempt to bind discourse to a central, singular meaning, to a univocal core, reflects *only* our need for practical exigency (a magnificent example of which is the pristine symbolizations of the discipline of mathematics which lead to ways of understanding each other that, ideally speaking, belie misunderstanding), and we become troubled only when explicitly *called upon* to produce a boundary around such a center. "If someone were to draw a sharp boundary I could not acknowledge it as the one that I too always wanted to draw, or had drawn in my mind. For I did not want to draw one at all. His concept can then be said to be not *the same* as mine, but akin to it. The kinship is just as undeniable as the difference" (Wittgenstein 1968, 33). And, as Wittgenstein further notes, I can *draw* boundaries or frontiers in such matters, but I cannot *give* such matters a boundary. In its lived, lively usage as a mathematical term, the term "higher" resonates in an untroublesome way, beyond the idealized frontiers that we can *draw,* but *cannot* give. In its lived usage, we are not troubled by the foreclosed alternatives of identity or difference, but can live in an untroubled way with the ambiguous and fluid *kinships* that language evokes. We were not compelled, in this Early Childhood Education class, to declared in the end that either 198 "really" *is* higher than 56, or that it *is not.* This question of whether it "really" is or is not put a demand on our discourse that we could not sustain if we desired to understand the discourse of children and the experiences they undergo in learning mathematics. Either of these foreclosed options belies the possibility of actually *teaching* mathematics to young children.

If we *begin* with and hold fast to the assumptions of exactness and literalism and abstractness and closure found in the discipline of mathematics, teaching young children mathematics becomes almost impossible. Understanding mathematics does not begin with literalism. In walking down the stairs, the two-year-old child's life is already pacing out mathematics. This child is already an initiate into "the community of mathematical users" and should not be considered an "outsider" because of the developmental inaccessibility of the abstract symbols that other members of this community can use. This was a difficult point for my students to understand. They deeply believed that mathematics *began* at the moment that its exactness, literalism, abstractness, and foreclosure began, and many of them hate mathematics for this very reason. Perhaps this hatred originates in the ways in which they were cut off from how mathematics already resonates in their own lives, their own experiences, their own understandings. Perhaps they were never allowed to be the vulnerable initiate, never allowed to explore and savor the fragile and delicate beginnings. Such beginnings *appear* to be left behind as mathematical understanding matures. But young children can still say to someone well versed in the discipline of mathematics that they are going to "count *up* to ten," and they will be understood *if* our attempts to understand do not peevishly wield the literalism of that discipline as a weapon of exclusion/exclusiveness (it is precisely this sense of "exclusiveness" that many of my student-teachers despise). Understanding "counting up to ten" is the sort of considerate and generous understanding that a *teacher* of mathematics to young children requires. If we listen carefully, we can hear that this young child is already one of us, already party to the community of mathematical users to which we also belong with him or her. He or she is one of us. He or she is a child.

Another topic that came up in this class was "symmetry." In the mathematics curriculum guides, it has become simply a vaguely incomprehensible theoretical term for an abstract notion of one-to-one correspondence of spatial relations. It has become sealed off in anonymously reproducible definitions that are clear but that do not touch anything outside of themselves. But symmetry also has to do with the slicing open of an orange and finding the pinwheel pattern in the midst of the aroma and sticky, bittersweet juice. It has to do with the fact that when children (or adults) clap their hands in steady rhythm out in front of themselves, they are not only involved in a similarity of meter (sym-metry, which itself interweaves with similarities of sound or sym-phonics, both of which come together in phenomena such as nursery rhymes—singing songs with young children is a deeply mathematical activity), but are displaying the symmetrical character of their bodies. Moreover, when we look at symmetrical objects, we tend spontaneously to put them "right in front of us," unconsciously centering the symmetrical object on our corporeal, symmetrical center. Symmetry is therefore *already* an unnamed feature of the Earth, and an unnamed feature of the young child's experience of the world. Children, prior to our earnest interventions as teachers, prior to the mandates of curricular accountability and reproducibility, have *already* sliced open an orange, or clapped their hands, and have therefore *already* lived in the unnamed, ambiguous presence of symmetry, division, fractions, circular patterning, and pattern subdivisions (both visual and auditory), one-to-one correspondence, numeration, and so on. If we forget this—if we forget that what we wish to teach is already at play in our language, in the Earth, and in both ours *and* the child's unnamed or unnoticed experiences—a whole image of teaching and learning emerges. It does not involve drawing out the deep resonances of language and experience in which the child already dwells (and in which we dwell with the child). It is in-struction, the making of knowing in the child from the outside; training, not educing. And a whole image of ourselves as teachers emerges. If we forget that we dwell *with* our children in the deep resonances of language and experience, we can forget our kinship with children. In becoming estranged from our kinship with children (with the fact that they are our "kind"), they can become our strange and silent objects, ones that have nothing of their own to say, ones we must now in-struct without feeling the need to listen to the unvoiced experiences they have already undergone.

Perhaps this is why, when one of my student-teachers recently realized that the topic of her lesson might be one with which the children were already vaguely familiar, she said "Wait, I'm not ready!" This is the beginning of a recognition that life goes on beyond our earnest intentions and actions as teachers. It is the first glimmerings of a precious realization so essential for student-teachers to undergo—that understanding erupts out of life itself and not simply as a response to our concerted acts of teaching and, therefore, that teaching must first and foremost attune itself to what is *already at work* in our lives and the lives of the children we teach.

Discovering this analogical play of language can initially have a disorienting effect. One begins to bump into the limits, contours, and interweavings of language itself (Wittgenstein 1968, 48), and the secure, familiar, but deadened ground of literalness and commonness begins to slip and move. One begins to realize that one's intentions and experiences are always already outstripped by the interplays of discourse itself. This may be why, when I noted that one of my students had slipped into the language of one number being "bigger" than the other, she responded with embarrassment, apology, and statements like "But I didn't *mean* that," as if she felt somehow culpable for the connections and implications of meaning which she had in fact *stumbled upon,* not *created* or *authored.* She discovered that language, just like children, always seems to be slipping out of our control, meaning more or less or different than we intended, interweaving in unanticipated ways with realms of meaning which we might have supposed to be separate and distinct. Our language and experience are *already* alive with interrelationships, kinships, and resemblances. The Earth is *already* alive with the very kinships our language speaks, even in the language of mathematics. Mathematics has an Earthen voice that lives beyond its deadly surface clarity. It is *right here,* at our fingertips.

With the particular example of higher and lower numbers, then, I did not want my students to solve or resolve the inherent ambiguity of this term "higher," but to recognize it as such. They have not discovered an error in their speech which they must, in principle, seek to remedy by rendering their speech univocal and unambiguous. Clearly, such rendering is not only possible (we *can* draw a boundary); it is often precisely what is required of us over the course of teaching your children. (It is, in fact, precisely the rendering required of the discipline of mathematics itself—mathematics *becomes* a self-enclosed discipline, oriented to literalism, and this is precisely where its beauty and its power lie.) But *this* recognition—that drawing univocal and unambiguous boundaries is often required, often called for—is quite different from the claim that there simply "are" such boundaries in all cases and that our task as teachers is always and everywhere to orient to such boundedness. *This* recognition makes literalism, univocity, and definition a *practical and localized choice* which salvages us from the absurdity of believing that we can "give" boundaries. It requires remaining attentive to what particular situations, particular speech, particular children, particular activities call for, what they require, rather than requiring of all situations, all speech, all children, and all activities that they live within the boundaries that we have already drawn. Since we can only draw boundaries and not give them, requiring teaching and learning to be always "living within boundaries" can only lead to frustration, absurdity, and despair. It can lead only to the belief that the breaching of boundaries is an act of violation, an act of willful misbehavior, when it in fact may be the spontaneous eruption of meaning that lies at the heart of language, the heart of education, *and the heart of mathematics* itself. The problem here is that there is no guaranteed "trick" for telling the difference, and that in taking the risk of allowing for the ambiguity and vitality of language, we at once risk violation, willful misbehavior, and the like. The game can get out of hand.

In attending to the interweaving ambiguities of language (rather than seeing such ambiguities as problems to be fixed), our delicate, irresolvable "kinship" with children can be brought to the fore. Such kinship is not a matter of being the same as children, but neither is it simply a matter of difference. Our kinship, both with the Earth and with each other, cannot be foreclosed by either of these extremes. Rather, it is a stunning, tensive *relationship* that will not resolve itself into either identity or difference. It is an *analogical relationship* (Jardine and Morgan 1988). As such, it holds fast to the unavoidable paradox that lies at the heart of education: children are "part of us but also apart from us" (Smith 1988b, 175). It also holds fast to the belief that the language of education—even mathematics education—cannot be caught in the immobility of literalism and univocity, for such *stasis* is essentially anti-educational, degenerative. Education—even mathematics education—must become celebratory of the ambiguity, generativity, and freedom of language, celebratory of its profound *humanity*.

Interlude: On Analogical Language

As in spinning a thread, we twist fibre on fibre. And the strength of the thread does not reside in the fact that some one fibre runs through its whole length, but in the overlapping of many fibres (Wittgenstein 1968, 32).

Don't say: "There *must* be something common" but *look and see* whether there is anything common to all.—For if you look at them you will not see something that is common to *all,* but similarities, relationships, and a whole series of them at that. To repeat: don't think, but look (66)!

We see a complicated network of similarities, overlapping and criss-crossing: sometimes overall similarities, sometimes similarities of detail. I can think of no better expression to characterize these similarities than "family resemblances" (31).

The difficulty with the mathematics curriculum is that it appears to not be conversant with anything outside of itself. It appears self-enclosed, complete, detached—it doesn't *speak,* it will not *answer.* What we have seen above is the beginnings of a type of movement or slippage in mathematical language equivalent to a movement beyond literalism and the potent lust for univocal definitions and boundaries. It begins to appear *dialogical, analogical,* instead of *monological.* It begins to appear as a form of *conversation with the Earth,* it begins to appear as something about which we might be able to have a conversation with children. As such, it becomes more unclear and difficult, *able* to withdraw into certainty, but forfeiting such withdrawal in favor of richness, Earthiness, humility.

Analogical thinking involves the development and exploration of likenesses, similarities, correspondences—parallels between worlds of discourse (*ana logos*). Such parallels resist the collapse of one realm of discourse into another, while also resisting the isolation of such realms. They involve, so to speak, the "conversation" or "dialogue" between such realms, a dialogue that sustains a "similarity-in-difference" (Norris-Clarke 1976, 66). Understanding and exploring an analogy, therefore, is not a matter of discovering some discursive, univocal term which makes both sides of the analogy *the same,* which collapses the "network of similarities, overlapping and criss-crossing" into literal terms which can be applied, *univoce* to both sides of the analogy. Rather, understanding analogies involves the exploration of the tension that is sustained between similarity and difference, a tension that *cannot, in principle, be discursively cashed out in just so many words.* Understanding an analogy is a matter of becoming party to the conversation between differing realms of discourse that the analogy opens up, "getting in on the conversation." The example used above of higher and lower numbers, then, was not to be taken literally, as if there were some univocal sense in which numerical quantity and notions of "higher" and "lower" were *identical.* But neither could it be maintained that they were simply *different.* Rather, they had a kinship which drew them together, yet kept them apart. The compelling and powerful character of analogies is not found in *solving* such kinships by resolving them into identity or difference—as if they were "problems" that found their resolution in translating them into literal, univocal discourse; as if they were "accidents" that needed to be "fixed"; as if they were merely "decorative" marginalia to the true text of language. Rather, they gain their power precisely in their *resistance to being "solved."* It is because of this resistance to being solved that potent analogies can always be readdressed. It is because of this resistance that they remain "alive." The "conversation" that the analogy opens up can always be taken up anew. Clearly, a true conversation requires the seeking out of some common ground of understanding. But such a "common ground" need not entail an *identity of place* which would make both parties to the conversation *the same.* If we are both the same, conversation would not be necessary. Rather, a true conversation seeks out a common ground (if we were both simply different, conversation would not be possible) in the midst of a recognition of difference—it is a dialogue, not a monologue. Because of this sustained tension between similarity and difference, analogies are essentially generative and provocative. They can always draw us in again and provoke us to reconsider what they have to offer.

W. Norris-Clarke points out another feature of analogical language that is of interest. Understanding an analogical term requires "running up and down the known *range* of cases to which it applies, by actually calling up the spectrum of *different* exemplifications, and then *catching the point*" (Norris-Clarke 1976, 68). This is a telling point. To understand an analogical term deeply and fully, we must cover the range of its exemplifications and catch the point of the analogy that cannot be said in so many words. This is not a matter of refusal to be specific. Rather, it is a matter of the *embodiment* of meaning in its diverse instantiations. The meaning of "higher" and "lower" is not separate from its instances, like some ghostly "idea" that could be offered up independently of its instances. It *is* (analogically speaking) the diverse instances. It is not hovering "above" them (another lovely analogical term). The whole image of understanding that analogical discourse requires of us is different from that of the understanding required by definition. "There is quite a bit of 'give,' 'flexibility,' indeterminacy, or vagueness right within the concept itself, with the result that the meaning remains essentially incomplete, so underdetermined that it cannot be clearly understood until further reference is made to some mode or modes of realization" (Norris-Clarke 1976, 67). To understand the analogical term, therefore, we must allow it "to expand to its full analogous breadth of illuminative meaning" (72). Moreover, "whenever it tries to become too precise, it contracts to become identical with just one of its modes and loses its analogical function" (69–70). However, recalling Wittgenstein, we can *draw* a boundary and induce such contraction, but we cannot *give* such identity, *produce* such a loss of analogical functioning once and for all. When we turn our backs and cease our diligence, the term "regenerates," or re-expands into its diverse instances.

A final feature of analogical thinking which is of interest here is the ability of an analogy to move us from what is known to what is unknown, from what is familiar to what is unfamiliar. The effect of such a movement is not simply that we shed light on an unfamiliar or novel aspect of our experience by relating it to more familiar aspects. Rather, analogies both "make the novel seem familiar by relating it to prior knowledge, [and] . . . make the familiar seem strange by viewing it from a new perspective" (Gick and Holyoak 1983, 1–2). It may be that "a venture into the alien is only possible on the basis of the familiar" (Gadamer 1977, 15), but such a venture does not leave the familiar untouched. It allows us to return to the familiar in a new way. After our exploration of mathematical language in this class, some students have said that they have *more* difficulty with language than before, *more* difficulty in finding comfort in the familiar. The work they did in this class was not meant to make their lives *easier*, but to begin to free them for the real difficulty, the real claim that language makes on us. Some have described how this experience has made them more careful in their language, more attentive to the lessons and themes that our language and the language of children have to offer. For some, however, it induced a sort of temporary paralysis, rendering them silent, speechless, fearful, in some sense, of the unvoiced and unintended implications of meaning that issue with every word. In the long run, this silence might be a good sign. It may be a recognition of how our language, our experiences, our *lives* as adults are always already deeply implicated in the lives of children. Out from under the noisy clatter of tricks and techniques they have mastered, they may have come upon their silent kinship with children. They have come upon an "original difficulty" (Caputo 1987) of education: living their lives in the presence of new life in our midst, (Smith 1988b) in the presence of those who are always and already our kin.

It is this final characteristic of analogical discourse—its relation to the familiar and its evocation of family resemblances and kinships—that serves as the focus for our concluding remarks.

Concluding Remarks

A friend of my son came to visit recently, and I told him about the huge pond in our neighbor's field. The spring runoff had created a slough about eight feet deep. After discussing that it would be over his head if he fell in, over my son's head, and even over *my* head, he asked, "If a hundred-year-old man stepped in it, would it be over *his* head too?" I answered, "Yes, it's *that* deep."

> Here it is difficult as it were to keep our heads up—to see that we must stick to the subjects of our every-day thinking, and not to go astray and imagine that we have to describe extreme subtleties. We feel as if we had to repair a torn spider's web with our fingers (Wittgenstein 1968, 50).
>
> The aspects of things that are most important to us are hidden because of their simplicity and familiarity. The real foundations of [our] inquiry do not strike [us] at all. Unless *that* fact has at sometime struck [us].—And this means: we fail to be struck by what, once seen, is most striking and powerful . . ..
>
> The most difficult learning is to come to know actually and to the very foundations what we already know. Such learning, with which we are here solely concerned, demands dwelling continually on what appears to be nearest to us (Heidegger 1977, 252).

Allowing ourselves to experience the freedom and generativity of language is at once endangering ourselves and the tranquility that our boundaries provide. This class was a painful experience for some students, and this is no surprise. For in language is embedded our understanding of ourselves, the world, and others; in language are embedded social, political, economic, and cultural tides that sway and pull and push us beyond our understanding, intent, or conscious concession. Disrupting our dwelling in language is at once disrupting ourselves, our self-understanding. For it is *we ourselves* who are at stake in the question of our dwelling in language. And, as teachers and parents who have a hand in what our children might come to live with, it is not only we ourselves, but the children all around us who are at stake. Passing on the deadened language of the mathematics curriculum guide is also passing on a self-understanding to children. In our hatred of the meaningless fantasies of mathematics, in our feelings of anxiety and incompetence, we pass on to children exactly this. Beginning to experience the liveliness and generativity of language might require letting go of deeply embedded threads of self-understanding that our schooling has produced.

Our explorations in this class did not end in a set of clear and distinct declarations that students could list and take away with them and apply to their teaching experience without risk or difficulty, without humility, without a sense of the delicate contingency involved in such matters. In our class, we never really seemed to "get to the point," but constantly encircled the topic, finding localized "nests" of meaning with no singular center. We found, with all the mathematics topics we covered, that myriad voices were involved. In the long run, the hope I had for this class was that students would become *conversant* with the mathematics of young children. It was an exercise in bringing mathematics down to Earth, re-enlivening it, and this entailed making it more difficult than the clarity of the curriculum guides suggests, but also more "at hand," more human as well.

Etymologically, "human," "humus," and "humility" are linked. Making mathematics seem more human entails the darkness of humus and Earthliness, with all its interweaving and intersecting threads. But it also entails a sense of something out of which things can grow, something alive or sustaining of life, something generative. This class, then, was an exercise in celebrating the generative and playful character of language, which goes on despite our desires to foreclose on this character. But, deeper than this, it was also an exercise in celebrating the generative and playful character of children, which goes on despite our desires to foreclose on *this* character. Savoring the irresolvable kinships of language is an analogue to savoring the irresolvable kinship we have with children. And teaching mathematics to young children requires welcoming these kinships and becoming attuned to them; it requires remembering how mathematics lives in and is regenerated through such kinships.

If the discipline of mathematics were as self-enclosed as it often announces itself to be (if it could *give* itself a boundary, and not merely *draw* one), there could never be any "new ones" among us; perfect self-enclosure is simply degenerative. It is precisely a loving, generative attention to these "new ones" that defines our special task as teachers. The problem with the literalism and exactness of the discipline of mathematics is that it can live in a forgetfulness of the young and become (pedagogically speaking) tragically self-enamored and self-enclosed. As teachers of young children, we are not allowed the luxury and comfort of self-absorption that issue from such self-enclosure.

How peculiar it still seems to consider all this as offering images of the thinking and language of mathematics. But when the three-year-old child announces that he or she is going to "count all the way up to ten," literal and exacting disciplined mathematicians will *already* understand what has been said, *even though* such understanding belies the literalism to which they may be professionally accustomed.

CHAPTER FOUR

"To Dwell with a Boundless Heart":
On the Integrated Curriculum and the
Recovery of the Earth (1990)

Prelude

I like to walk alone on country paths, rice plants and wild grasses on both sides, putting each foot down on the earth in mindfulness, knowing that I walk on the wondrous earth. In such moments, existence is . . . miraculous and mysterious. People usually consider walking on water or in thin air a miracle. But I think the real miracle is . . . to walk on earth. Every day we are engaged in a miracle which we don't even recognize: a blue sky, white clouds, green miracle (Hahn 1986, 12).

I began teaching my undergraduate Early Childhood Education (ECE) class this year with handing my students a blank piece of paper and instructing them to write down as many possible ways in which the paper could be used to demonstrate, illustrate, or teach features of the various curriculum areas.

Their ideas began as expected, with possibilities such as writing on it, painting or drawing on it, reading from it, folding it and making shapes, questions of where paper comes from, how it is made and used, and so on. But, in the midst of this exercise, there was a striking advent for this class. Once they moved to questions of how it was made, one student suggested that you could talk about trees and still remain "linked up" with this paper, still remain, so to speak, "on topic." Once this shift of focus occurred, what began was a giddy onrush of sun and soil and water and logging and chainsaws and gasoline and refineries. That is to say, because of this serendipitous "turn" of attention, suddenly and unexpectedly, *everything* came to be co-present with this paper, everything seemed to nestle around it. Some topics seemed quite "close" to this paper, while others were quite distant, at the ends of long and tenuous tendrils of interconnection. Some connections were obvious and immediate, some connections were "stretched," but *nothing* was absent altogether.

One striking feature of this class was that we seemed to go beyond a mere "mental exercise," glimpsing something about the world and our experience of the world, a previously unnoticed interconnectedness of things hidden beneath the surface analytic assumptions of difference and separateness that are so commonplace and that guide much curricular thinking.

> With the interdependence of all things or "interbeing," cause and effect are no longer perceived as linear, but as a net, not a two-dimensional one, but a system of countless nets interwoven in all directions in a multidimensional space (Hahn 1988, 64).
>
> All things in the world are linked together, one way or the other. Not a single thing comes into being without some relationship to every other thing (Nishitani 1982, 149).
>
> Even the very tiniest thing, to the extent that it "is," displays in its act of being the whole web of circuminsessional interpenetration that links all things together (Nishitani 1982, 150).

As we proceeded with reflections on this exercise, we realized that *any* object could have been utilized for this demonstration, *any* object could have been "drawn into the center" in a way that all other things organize themselves around this center. With any object, everything else seems to come forward as implicated in this object, but there is no "special" object in this implication which has a privileged status as center.

> The universe is a dynamic fabric of interdependent events in which none is the fundamental entity (Hahn 1988, 70).

Although it is clear that, for example, a piece of blank paper "lends itself" to curricular matters that are proximal to it (e.g., writing, drawing, questions of how it is made, etc.), pulling out this piece of paper tugs at the whole fabric of things without exception. Paradoxically put, then, *every* object is a unique center around which all others can be gathered *while at the same time* that very object rests on the periphery of all others, proximal to some, distant to others.

To say *that a thing is not itself* means that, while continuing to be itself, it is in the home-ground of everything else. Figuratively speaking, its roots reach across into the ground of all other things and help to hold them up and keep them standing. It serves as a constitutive element of their being so that they can be what they are, and thus provides an ingredient of their being. *That a thing is itself* means that all other things, while continuing to be themselves, are in the home-ground of that thing; that precisely when a thing is on its own home-ground, everything else is there too; that the roots of every thing spread across into its home-ground. This way that everything has being on the home-ground of everything else, without ceasing to be on its own home-ground, means that the being of each thing is held up, kept standing, and made to be what it is by means of the being of all other things; or, put the other way around, that each thing holds up the being of every other thing, keeps it standing, and makes it what it is (Nishitani 1982, 149).

There is no singular center that will resolve this paradox, that a thing is, so to speak, *not itself* (i.e., it *is* only in relation to all other things and therefore summons up all those things that it is *not* in order to *be* itself) *while being itself.* If it were not for trees and sun and sky and water, there would *be* no paper, and to fully understand what this piece of paper is in an integral way requires bringing forth this paradoxical, interweaving indebtedness. The name for this paradoxical, interweaving indebtedness is the Earth.

From here, our class moved on to a discussion of the question of the nature and assumptions of an integrated curriculum and the appropriateness of such a curriculum for Early Childhood Education.

The Recovery of the Earth

The notion of the "integrated curriculum" is becoming common currency in Early Childhood Education in Canada, and the articulation of this concept across grades K-6 is beginning in some circles. What seems to be missing in many current formulations of this notion is any deep sense of the *difference* it makes in our lives and the lives of children. Is it simply a new slogan which will become exhausted and empty as have so many others in the consumptive flurry in education for the newest and the latest? Or does it speak of something new, something vital and generative in the field of education? I believe that it is potentially the latter.

But this potential is a difficult one to assess and address. The exercise my students did in our class was a momentarily enjoyable one, but it is also one whose giddy insight is difficult to sustain. My students did report that they "glimpsed" something about the notion of integration in the curriculum, but it was almost impossible to sustain this glimpse and cash it out as something *practicable*. It was difficult to lay out in front of us as a set of propositions or formulae, not because of the complexity of this task or its arduous nature, but because what we were glimpsing was precisely not an object for our perusal or an objective set of relationships that we can set before us. Rather, we were glimpsing the way in which the Earth is our abode, our dwelling, and how our lives as teachers are an integral part of this dwelling.

It must be said, however, the notion of an integrated curriculum became a painful one for some students as they began to confront the fossilized residues and assumptions of their own schooling and, more pointedly, as they began practice teaching in situations of profound dis-integration. This seemingly innocent and playful exercise we conducted did not make matters easier or clearer, nor did it make questions of applicability more simple and straightforward. It made things worse. Underlying this difficulty are questions regarding images of our lives and the lives of our children that both sustain and "ground" the notion of the integrated curriculum. Something archaic and delicate and difficult needs to be recovered in order for the integrated curriculum to have any deep sense.

> The unnoticeable law of the earth preserves the earth in the sufficiency of the emerging and perishing of all things in the allotted sphere of the possible which everything follows, and yet nothing knows. The birch tree never oversteps its possibility. It is [human] will which drives the earth beyond the sphere of its possibility into such things which are no longer a possibility and are thus the impossible. It is one thing to just use the earth, another to receive the blessing of the earth and to become at home in the law of this reception in order to shepherd the mystery . . . and watch over the inviolability of the possible (Heidegger 1936–46, 1987, 109).

If we begin to unearth the notion of the integrated curriculum, it begins to disrupt deeply held beliefs and images of understanding, self-understanding and mutual understanding, pointing to a sense of "inter-relatedness," "inter-dependency," or "inter-connectedness," which is belied by our analytic, definitional, and frequently dis-integrative approaches to educational phenomena. It also belies the desire to finalize, control, master, and foreclose upon vital curricular issues. It puts into question desires we may have, as educational theorists and practitioners, to get the curriculum "right," "straightened out" once and for all, for such desires require a basically disintegrative, analytic act aimed at rendering education a closed question, aimed at rendering human life lifelessly "objective" under the glare of knowledge-as-*stasis*.

Integration leads to glimpses of a truly lived curriculum, a true *curriculum vitae,* one that exudes the generativity, movement, liveliness, and difficulty that lies at the heart of living our lives, as educators, in the presence of new life in our midst (Smith 1988b, 175), in the presence of children. A truly integrated curriculum involves the ambiguous and difficult ways in which our lives are intertwined with children—the irresolvable paradox of children "being part of us but also apart from us" (Smith 1988b, 175)—and the ways in which our lives together with children are interwoven with the life of the Earth. It is *this* "integer," *this* "whole," *this* "integrity" that the integrated curriculum voices.

There is, near the roots of the notion of the integrated curriculum, a strikingly simple image of education: "The essence of education is natality, the fact that human beings are *born* into the world" (Arendt 1969, 174). And "to preserve the world against the mortality of its creators and inhabitants, it must be constantly set right anew. The problem is simply to educate in such a way that setting-right remains actually possible, even though it can, of course, never be assured" (192). Education, in this image, has to do with our fundamental orientation to the fact of natality and, therefore, our fundamental orientation and openness to the future. Although it tends to ceaseless proliferation of longer and longer lists, guides, schedules, and agenda, education, at its heart, cannot be caught in the *stasis* that such a tendency requires and desires in the end. Rather, education is *ek-static,* a movement beyond what already is, a reaching out to the new life around us in a way that keeps open the possibility "that the people of this precious Earth . . . may live" (Fox 1983, 9).

The central claim of this paper is this: the integrated curriculum is, at its roots, more than a matter of the inter-relations between curriculum areas or subject matters. It is an ecological and spiritual matter, involving images of our place and the place of our children on "this precious Earth." It raises the question of how we are to understand that we are people *of* this precious

Earth, caught up in its potentialities and possibilities. It raises the question of how the deep and moist interweavings and integrity of the Earth is both an original constraint on our lives, but also an original blessing (Fox 1983, 83), an original freedom, the overstepping of which pushes the Earth beyond what is possible for it to sustain. And, in the end, the integrated curriculum requires a deep reflection on our desires to dis-integrate children's curricular experiences in the name of manageability, easy of instructional design or territorial notions of the separateness and uniqueness of subject-matter specializations.

As an ecological and spiritual matter, the notion of the integrated curriculum involves disturbing, even horrifying questions as to whether this basic fact of natality that springs from this original blessing of the Earth and that lies at the heart of education can, in all its ecological delicacy, be taken for granted any longer. We could never and cannot now *assure* our children an Earth on which life can go on, an Earth on which "setting-right is actually possible," for such assurances are quite literally beyond us. The horror is that degenerative, dis-integrative, and consumptive images of human life and the bringing forth of human life (*educare*) may be assuring the opposite. The true horror is that *this* assurance is precisely *not* beyond us *even if* we choose to ignore it and live, educate, and proliferate educational theories, research, and practices as if the Earth does not matter (see Devall and Sessions 1985), as if, therefore, the continuation of human life were not an educational concern.

Ignoring the ecological and spiritual consequences and character of the integrated curriculum plunges education into a peculiar paradox, an "impossibility." We are able to diligently pursue ways in which the mathematics, science, social studies, and language-arts curricula may be taught without ever considering whether such diligence, such curricula, and such teaching operate in a way that is consistent with the continued existence of an Earth on which such knowledge may be brought forth. *Educare*—"bringing forth"—is understood, so to speak, "from the neck up," as if it just happens in the head, as if it were just a matter of effective teaching and affected learning, requiring no *real* place, no *real* space in which it occurs.

Such a strangulated approach to education forgets that it is not accumulated curricular knowledge that we most deeply offer our children in educating them. It is not their epistemic excellence or their mastery of requisite skills or their grade point average that matters most fundamentally, but quite literally their ability to live, their ability to *be* on an Earth that will sustain their lives. If we begin to take the roots of the integrated curriculum seriously and begin to heed what it requires of us as educators, we must educate and we must understand the curriculum in ways that will sustain the possibility that all our efforts, and all the efforts of our children, and all these matters of so much concern in educational theory and practice will not be suddenly trivialized. A thorough "grounding" in mathematics is of little use if that knowledge is understood in such a way that there is no longer any *real* ground that is safe to walk. Mathematics must become *earthen* in how it is understood, how it is taught, and how it is "grounded." This does not require tampering with the indigenous sense and operational character of mathematics, but it *does* require that we face the fact that actually producing, sustaining, savoring, and passing on such knowledge requires something more than this sense and character—it requires an Earth.

We can, as Wittgenstein put it, *draw* a boundary around, for example, the mathematics curriculum (and it is, on occasion, completely appropriate to do so) but we cannot *give it a boundary* (Wittgenstein 1968) that could prevent it from intertwining with our lives and the life of the Earth. One cannot sensibly aspire to well-bound and defined and circumscribed images of knowledge and of being educated if those images belie the existence of the actual breath required to pronounce that aspiration. No matter how careful we are in our drawing of boundaries, mathematics interweaves with the fabric of the Earth. My love of mathematics, then, must remain Earthbound—it must remain a love not only of its indigenous and articulate beauty, but of the actual conditions under which I and the children I teach may live to do it. And it is *this* love and understanding of mathematics that I must pass on to the children I teach. Within an integrated curriculum, *this* is what it means to sensibly say that I "teach mathematics." It means teaching a love of the Earth on which the teaching and learning and savoring of mathematics is actually possible.

Admittedly, this paradox always seems to be overstated and rather hysterical. But *this* is where the integrated curriculum really begins to hit home as something that goes beyond precious notions of the relations between curricular subject areas and impotent epistemological notions of "relevance to the child's life." It is concerned with the knowledge—perhaps we must say the "wisdom," even if we find such notions vaguely embarrassing, antiquated, unrigorous, or unclear—that we must pass on to our children so that life on Earth can go on. It is concerned with an image of knowledge that our children can *live* with, with relevance to the child's *life*. It is precisely *not* a specialized curricular topic such as "ecological studies" or "environmental studies," for such specialization unwittingly pretends that the Earth is not underfoot *no matter what*. The Earth and its continued existence is not a specialized topic among others as if these others were exempt. The integrated curriculum, understood as an ecological and spiritual matter, throws back in our faces any such presumptions of being exempt.

The curriculum as an integrated curriculum cannot be considered merely conceptually, *as if* we and our children are ghostly, objective purveyors of the Earth, and not fully *human*, full of *humus*, fully embedded in the life of the Earth, fully indebted. Although such intellectual celibacy produces beautiful and seductive educational edifices—new theories, new guidelines, more and more complex educational agenda, and longer and longer lists of strategic teacher intervention procedures—we cannot *live* with such edifications unless they are somehow brought down to Earth, "grounded," not only in an epistemological sense, but in a moist, fleshy sense of given earthliness, given *humus*, made human. To do this, we may have to forfeit some of our precious clarity and distinctness (this was a problem my students faced in our class—things became more difficult, blurred, but also richer and more "down to earth" than the theories with which they are often inundated). We may have to admit that the continued existence of our lives and the lives of our children contain an Earthen darkness and difficulty—an Earthen life—that we have heretofore fantasized out of curricular existence. The integrated curriculum, understood out from under these celibate fantasies, requires a recovery of the delicate, interweaving, and intertwining *humus* of a *curriculum vitae;* it requires a recovery of the Earth.

On the Language of Curricular Discourse
and "Teacher Responsibility"

One of the difficulties in writing (and, I suspect, reading) this paper should be explicitly admitted. Attempting to write about the integrated curriculum as an ecological and spiritual matter is at once a struggle with the language of curricular theorizing itself.

The language and tenor of educational theory and practice has, in many circles, taken on the interests, hopes, terminology, techniques, hesitancy, carefulness, and canons of objectivity of scientific discourse. Those forms of educational theorizing that do not take on such language and tenor often fall prey to all-too-easy caricatures of its alternative—subjectivism, anthropomorphism, individualism, experientialism, narrative, personal accounts, unrigorous and undisciplined swooning, and the like.

It would be preferable to simply announce that one is going to sidestep this lover's quarrel but this is difficult to do, for the language of an integral alternative is not readily available. The integrated curriculum requires a "whole" language, but it cannot be caught in the all-too-frequent "profligacy of self-annunciation" (Smith 1988c, 247) that infects some proponents and popular conceptions of "whole language theory."

One of the claims of an ecological understanding is that life on Earth involves a multitude of different interweaving and intersecting voices, of which the human voice is but one among many (and, of special interest in education, of which the *adult's* voice is but one of these). Living with the richness and difficulty of this multitude of voices and speaking out from the midst of it is *part of the phenomenon of the integrated curriculum.* The struggle with curricular language—including the problem of "overheated prose" evident here and elsewhere—is not an *accident* which one must first rectify before inquiry begins, but is precisely what must be recovered in the recovery of the Earth. The integrated curriculum does not require or allow the reduction of this multitude of voices to a single voice (univocity, evident in the desire to reduce all voices to a unique, single center, evident in literalism) but neither does it require or allow the closing off of different voices in their difference (equivocity, evident in the bound character of separate curriculum guides, specializations, etc.). The question of how life on "this precious Earth" can go on is a question of how *the conversation between different voices* can go on.

Such a conversation requires more than just speaking or brazen self-annunciation. It requires listening, attending, and attunement to other voices. And, more disturbingly put, it requires more than the numbing and light-headed enthusiasm and "positiveness" that can often accompany teaching at the early grades. It may require attending to "the negatives, that is, the silences, the blockages, the unspeakables of life" (Smith 1988c, 247). It may require, albeit in an developmentally appropriate way, that we tell our children the truth. It may require that we listen to our children or to the voice of the Earth, even if such listening is difficult, perhaps painful, perhaps disruptive of the clear and distinct boundaries we have set for ourselves and our children.

But then, on the reverse side of this coin, there is also a sort of playful simplicity to language that can issue from considering the "conversational" or "dialogical" nature of the integrated curriculum in Early Childhood Education. As potential educators of young children, my students have the excuse to *re-experience the world.* The children they will be teaching are in the process of learning what they, as adults, now take for granted, and, as teachers, they can allow their experience of the world to become new again. They can begin to have anew a conversation with the Earth, to notice anew what has gone unnoticed under the rubric of familiarity, ordinariness, and the like.

I expect that some of my students have been schooled to believe that understanding the young child's experience and curriculum must be something esoteric, unfamiliar, unordinary. Some were expecting long and involved lists of peculiar theoretical characteristics and articulations. If the integrated curriculum is to be understood as an ecological and spiritual matter, however, it must cash itself out right here, in the re-generativity and re-engagement of the simplest of events, right here at our fingertips. *This* water ring on the table left by my glass can be seen as just an incidental event, ignorable, worthy of indifference. But it is also the occasion to become enchanted again. It embodies whole realms of experience, vast complexities and interrelations to be explored—heat, cold, water, water vapor, humidifiers, evaporation, condensation, clouds, rain, snow, and, I suspect, that piece of paper with which we began. The discovery of such an example is not the result of a vast reservoir of theoretical experience, but a sort of attendance and attunement to the minutia of our lives and a forfeiting of our schooled tendency to deaden language and experience by taking the boundaries we have drawn too literally, as "closed boundaries" that know no play, no interplay with what is around them. It is not just a piece of "information" about water rings that one possesses in this instance. It is, rather, a sort of dis-possession, a letting go rather than a grasping; deeply understanding the integrated curriculum becomes a matter of "self-transcendence" (Phenix 1975, 323).

The struggle my students confronted in our play with the piece of paper was, in part, a struggle with language. Some students adamantly began this exercise with declarations like "come on, it's just a piece of paper," demonstrating how familiarity can breed contempt and a sort of ungenerative *stasis,* a desire to "hold on," to the boundaries already laid out. They could not easily become "conversant" with this paper, because they believed, in essence, that the last word had already been said, that there was nothing really left to say—"it's just a piece of paper." The danger with such a holding on to boundaries is, of course, this: boundaries are meant to keep others (other meanings, other interpretations, other understandings) out. *And it is precisely "others" who we are dealing with in educating young children.*

The danger with such a holding on to boundaries is that it can cash itself out as a contempt for children, a contempt for their difference. If we begin a career in education with the belief that there is really nothing left to say, that the conversation is closed, that the boundaries have already been given to things, that we already understand, we begin unwittingly with the degenerative belief that the heart of education—the basic fact of natality—is simply a mistake to be corrected through our efforts, that the *difference* of children is a problem to be solved. And I suspect we have all lived in classrooms where such deadliness holds sway.

The importance of a trivial example such as this water ring is that young children have *already* experienced a "sweating glass" of ice water, a steamed up window, a scraped windshield in the winter. The interweaving possibilities and potentialities of the Earth are right at their fingertips already. The integrated curriculum, then, includes an acknowledgment of the fact that the child's experience of the world is *already* fully interwoven with our lives and with the life of the Earth, already "integral." It requires that instruction begin with and savor this "already," and that student-teachers develop a deep love for the generativity and liveliness of language itself. The contemptuous, deadened familiarity with the world with which some student-teachers begin their education can lead them to believe that such playfulness and generativity is simply a violation of boundaries which must be corrected (perhaps this is why so much of teacher education seems to be fixated on issues of "discipline" and "management"). What must happen in a turn to an integrated curriculum is that such familiarity must be deeply disrupted. But this must be done, not in order to turn away from the familiar to some unearthly discourse. Rather, such disruption allows us to begin a recovery of a deep sense of the familial, a deep sense of our inviolable "kinship" with children and with the Earth.

The integrated curriculum resists the degenerative tendency in education. It does *not* require an image of education as involving no "discipline." It does not involve education-as-chaos any more than believing that the Earth itself is chaotic without our concerted, authoritative intervention. Rather, it involves learning to live with, and learning to take educational advantage of the "discipline" and "organization" *originating from things themselves* and originating from children's spontaneous interest in the world, their *inter esse,* their "being in the middle of things." Once children's "interests" are understood as having a certain inviolable integrity, and once the contours and textures of what is being taught have been savored and explored by the teacher (i.e., once the teacher deeply *understands* his/her material, and, dare I say, learns to love the world again), taking educational advantage of such interests by drawing children into these contours and textures will help *prevent* those "discipline problems" issuing from misunderstanding children and not deeply understanding one's material. The teacher, in such an instance, becomes a facilitator, a provocateur, and, one hopes, a joyous *example* of a loving interest in children and in the contours and textures of the Earth.

Clearly and admittedly, this sounds desperately naive, for the teacher is the one who is *responsible* for his or her classroom and *responsible* for his or her children's education. An integrated curriculum certainly requires responsibility. However, such responsibility must be linked up with *precisely* such a loving interest in the Earth, including a loving interest in children. Our adult responsibility for and authority "over" children is at once a responsibility to the Earth on which we dwell *with* children. Teaching is, in part, an introduction of children to the authority of the Earth itself, an authority to which even *our* authority as adults is secondary. This is simply another way of saying that, no matter how loud our declarations or brazen our "authority," water runs downhill, human blood is warm, 2+2=4, and this piece of paper requires sun and sky and water. In an integrated curriculum, then, it is this deeper authority that requires our obedience and the obedience of children. And obedience, in the face of the archaic authority of the Earth, loses its moralistic character and can be finally heard again in its origin—*ab audire,* to listen, to attend, to be attuned.

> It is as if young people ask for, above all else, not only a genuine responsiveness from their elders, but also a certain direct authenticity, a sense of that deep human resonance so easily suppressed under the smooth human-relations jargon teachers typically learn in college. Young people want to know if, under the cool and calm of efficient teaching and excellent time-on-task ratios, life itself has a chance, or whether the surface is all there is (Smith 1988b, 175).

Concluding Remarks I: "To Dwell with a Boundless Heart"

The title of this paper voices how we might understand ourselves, not an *exception* to this interweaving indebtedness and inter-relatedness to the Earth, but as an *instance* of it. To dwell with a boundless heart is to understand "the self in its original countenance" (Nishitani 1982, 91) as delicately interwoven in this very Earthly fabric in which we found woven all things, including the children we teach. We can *draw* boundaries around ourselves (and it is very often appropriate to do so), but we cannot *give* ourselves boundaries without believing in the impossible—that our lives can go on, that we can *be,* without an ongoing conversation with "this precious Earth," one that includes our "knowledge" of it, but also includes our breathing of it.

> The self is here at the home-ground of all things. It is itself a home-ground where everything becomes manifest as what it is, where all things are assembled together into a "world." This must be a standpoint where one sees one's own self in all

things, in living things, in hills and rivers, towns and hamlets, tiles and stones, and loves these things "as oneself" (Nishitani 1982, 280–81).

There is a sense, then, in which this interrelatedness of things underlying the integrated curriculum requires seeing every action as an action on behalf of all, every thing speaking on behalf of all things. This concluding remark ends with a vignette.

Following a recently heavy oil spill off the coast of Washington State, my six-year-old son and I were watching the C.B.C. news. We saw film footage of an oil-covered duck clamoring up on to a beach on the west coast of Vancouver Island. With each panicked lunge, its wing-tips remained adhered to the slickened beach. This was followed by pictures of dead water fowl being shovelled up and put into green garbage bags.

We have all seen these sorts of scenes before, perhaps all too often. And I have often felt rage or sadness, or simply turned the damn thing off. But when my son turned to me and asked me to help him understand what he was seeing, I felt something new. I felt humiliated.

Even though it is all too easy to over-romanticize and anthropomorphize this point, I suddenly felt my own humanness as rooted in the same soil as this creature, my own "humusness." In attempting to understand this event and trying to help my son understand, I felt a sudden humility in that attempt, as if our understanding, our conversation, had to be brought down to Earth, humiliated in the proper sense. My son and I had to face our own indebtedness to this creature, to this oil, to this water, to this sand, to these scenes, to the power of these broadcast images, linking us to the production of this power, to its use, to the demands for it, to our demands for powering fuels, and then back to this oil, to this water, to this sand, to these scenes.

Watching and attempting to face these images, to make sense of them, produced the need for the very sort of fuel that was now killing this creature. It was as if it was undergoing this pain on our behalf. To tell my son of oil tankers and accidents and clean-up efforts no longer seemed like the whole truth. Such telling seemed like a dis-integrated curriculum guide version of the truth, where the pain and indebtedness are laid out anonymously before us to either peruse or ignore at our leisure. I had to try to tell him that we *cannot* "turn the damn thing off" by just switching off power to the television set. Not facing these images, turning *them* off, does not dispel our debt.

This realization was made all the more difficult when my son and I watched a video later that day, and a particular speech, in another context, in another place and time, hit too close to home, making the early scenes of ducks and oil and death, the earlier thoughts of indebtedness and humiliation, even more deeply unforgettable:

> I am asking you to fight—to fight against their anger, not to provoke it. We will not strike a blow, but we will receive them and through our pain we will make them see their injustice. It will hurt, but we cannot lose. They may torture me, break my bones, even kill me. But eventually, even in my death, they will see their injustice and they will stop.

Clearly, in quoting this film-version speech by Mahatma Gandhi, I am guilty of a sort of gross anthropomorphism, but the evoking of the roots of the integrated curriculum as an ecological and spiritual matter does require a deeper, different response than those I have become accustomed to as an "educational theorist." It requires a language of implication, of debt, of interrelation, a language that does not allow indifference, that does not allow us to "turn the damn thing off." Perhaps the language that allows us to call this unfortunate incident a separate event, discrete from curricular matters in the business of education—perhaps *that* is the truly anthropocentric language, believing as it does that the boundaries it *draws* it actually *gives*.

Nearing the finish of this paper, news of the Valdez oil spill in Alaska . . .

Concluding Remarks II: "This" Piece of Paper

It is too easy to become swept up in the happy interrelations of sun and sky and clouds and rain that nestle in this piece of paper.

This piece of paper, this very one that I am writing on, this very one that you are now reading, may be the one the bleaching of which produced the dioxin that may have already given Eric, my six-year old son, cancer.

Overstatement? Yes, perhaps. But, as a colleague once said to me, we will be responsible to our children for the questions that we do not ask. This—considering the life of my son Eric and what I will say when the questions come—is the real topic and the real cost of the work I do. It is the real sense in which the curriculum becomes a *curriculum vitae,* having to do with the course of our lives as they are actually lived.

We, in education, may be especially responsible for the questions we do not ask, standing as we do at the cusp of the emergence of new life in our midst, *able* to bring forth these questions, but perhaps unwilling to speak our real indebtedness to "this precious Earth" without embarrassment. The integrated curriculum has, at its roots, the potential to open up these questions we may have thus far refused to ask. Turning away from these questions may involve abandoning our children to an all too certain future.

CHAPTER FIVE

"A Bell Ringing in the Empty Sky" (1993)

I

Whereas a church spire inspires me to lift up my eyes to the heavens above, entering a tea room inspires in me something different. The entrance to the ceremonial room, by the very way it is built, urges me to incline my body and to bow, bringing me closer to the earth whose textured layers of humus allow buds of tea trees to leaf. The savoring of the tea allows me to touch again this earth that cradles and nourishes both my body and soul. During the Tea Ceremony, I come to respect the fullness of silence, and I become aware of how silently I participate in the constituting of that silence. And in that silence, I experience being-one-with-the-earth (Aoki 1987, 67).

The title of this paper is taken from a piece of Japanese music performed unaccompanied on a wooden flute called a shakuhachi, performed by Goro Yamaguichi. The piece, whose Japanese title is *Koku Reibo,* was originally composed by Kyochiku, a Zen priest. It is said to have come to him in a dream during which he attained enlightenment—that moist, green, Earthy enlightenment that de-spirited Western aspirations seem only to look down upon.

This piece of music and its evocative title call up a sense of openness and spaciousness; images of echoing and resonances; images of the time and the quiet needed for a conversation to go on; images of the space for the natural affections of speech, its kindness, its interweaving and living kinships or "family resemblances" to come forth.

I suspect that it is evocative partially as a response to the clangor in which many of us live—the urban, cluttered, and noisy abode out of which our curriculum theorizing often arises as a hurried response to our shortened breath. Our aspirations are strangulated into an unearthly discourse that must always have something new and notable to say at a moment's notice. We live, it seems, in a relentless proliferation of speech which leaves us neither time nor occasion nor place for a settling word, for quietly shepherding the mystery of our earthliness which is no mystery at all, which is right at our fingertips no matter what. And, more horrifying, it is we *in education* who are so caught up in this senseless roar and rush. Silence has become stupidity; taking time or hesitating before you speak has become a sign of incompetence or a lack of mastery in certain specifiable "communication skills"; speaking has become self-declaration; failing to declare oneself has become weakness; speaking with one's own voice has become gastric self-reporting. The living Word, and the silences and spaces it requires to hear its calling, is succumbing to the telegraphed stutters of televangelistic discourse.

What is so very horrifying is that, whether we intend it or not, we *are* what we deeply hope our children will become. Recently, a kindergarten teacher in the Rocky View School District in Alberta was fired by the parents advisory board because, at the end of kindergarten, not all of the children could read. And I find myself tongue-clucking at the pedagogical inappropriateness of such actions while rushing to meet publication schedules, requiring my student-teachers to compose hurried-yet-reflective journal entries about their teaching practice, and contemplating the need to write a book in order to secure career advancement. My son Eric, like any child, inhales my aspirations while breathing his own breath. This air, this atmosphere, is the real work (Snyder 1980) I do no matter what I say.

This certainly veers too close to home—that we may be offering our children an Earth on which only such ever self-present exhaustion, noise, and hurry is not only *possible,* but rather widely desirable. Such offerings are full of the "sadness, absurdity and despair" (Merton 1972, 297) that issues from being deeply "unearthed," deeply "unhuman," deeply out of touch with our humus. We have become schooled into aspirations that draw our eyes upward, or, more horrifyingly, inward into the seductive Cartesian allure of self-presence, self-stimulation, and self-annunciation (Smith 1988c, 247). In *Nausea,* Jean-Paul Sartre's fingertips are no longer flesh of the same Earth as the tree he touches; they are no longer flesh at all. His fingertips are only his reflective self-awareness of that touching. As Gary Snyder explains in an interview with Geneson:

> Geneson: So when Sartre . . . goes to the tree, touches the tree trunk and says, "I feel in an absurd position—I cannot break through my skin to get in touch with this bark, which is outside me," the Japanese poet would say . . .?
>
> Snyder: Sartre is confessing the sickness of the West. At least he is honest. The [poet] will say, "But there are ways to do it, my friend. It's no big deal." It's no big deal, especially if you get attuned to that possibility from early in life (Snyder 1980, 67).

Something is awakened here that is beyond the nightmare of self-presence and its ensuing exhaustion. It is a "call to be mindful of our rootedness in earthy experiences" (Aoki 1987, 67). But more: such a call can best be heard "if you get attuned to that possibility from early in life." Attunement. But also pedagogy. The possibility of touching the Earth, this attunement, is rooted (perhaps also uprooted) early in life.

Mindfulness of our rootedness in Earthy experiences is a breakthrough to the belonging-together of things that goes on without us, without our doing. It is a realization of the deep, Earthly collectivity of things that is not of our own making, wanting, or doing. In one formulation, hermeneutics seems to verge near this: the hermeneutic project is concerned with, "not what we do or what we ought to do, but what happens to us over and above our wanting and doing" (Gadamer 1989, xviii). But this hermeneutic project still remains a matter of Eurocentric enculturation, lacking the scent and the fragrance and the fleshy intersections of an Earth that "happens" (even if it *doesn't* happen "to us") and to which we are indebted in silent ways that speak neither Greek nor German. In mindfulness of this silent collectivity of the Earth, there is an archaic debt at work, the debt of breath and blood and sun and soil and sky, and all those hopelessly naive things the forgetting of which threatens to suddenly and violently trivialize our urbane theorizing with unspeakable ecological events that we, in all our earnestness, cannot outrun or sidestep. We pay homage to the "grounds" of our talk by evoking the thrilling names and concepts scattered over the history of phenomenology, hermeneutics, critical theory, deconstruction and, whatever is named as next with the loudest voice while forgetting the real ground right under our feet, the real breath we inhale to speak this homage, the real breath we hold in breathless anticipation for the latest news.

II

"A Bell Ringing in the Empty Sky": Both title and music bring out a
sense both of the exquisite solitariness of nature, and of the vibrant and
crawling interrelatedness and resonances of the Earth. It brings out *both of
these*—both the interrelatedness and the solitariness, both the intertwining
kinship of things and the uniqueness, individuality, and utter irreplaceability
of every thing. For this bell *is* in an empty sky, solitary and unsurrounded;
and it is *ringing out,* echoing and resonating with all things. It strikes an
image of interdependency and relatedness that *issues from* exquisiteness and
is not at odds with it. It strikes against an old image, rooted in Aristotle and
coined all too clearly by Descartes: "a substance is that which requires
nothing but itself in order to exist" (c.1640, 1955, 275). This is a perfect
segue into the unearthly loneliness we now are facing and are solemnly
passing on to our children under the guise of education and the individual,
manic pursuit of excellence-as-self-absorption—and with this, all the little
built-in panics and terrors that come with the blind rush to be up-to-date,
ahead of the game, or, as phallocentrism would have it, the potent desire to
end up "on top."

A bell ringing in the empty sky says this: any thing requires *everything
else* in order to exist. Each thing stands before us on behalf of all things, as
the absolute center of all things, *and* at the periphery of all things. It is an
absolute, irreplaceable, exquisite existence because it is empty (Japanese,
Ku; Sanskrit, *sunyata*) of self-existence. This paper from which you are
reading does not simply announce itself, announce what it "is." It is not
exquisitely this piece of paper because it requires nothing but itself in order
to exist; it is not a *substance* in the Scholastic sense. Rather, it is what it is
because it is what it is not—it announces sun and sky and earth and water
and trees and loggers and the meals they eat and chainsaws and gasoline and
pulp and the dioxin produced by the bleaching of this paper and the effluent
and the poisoned fish near pulp mills, and the cancer and the pain and the
death and the sorrow and the tears, and the Earth and the trees growing up
out of it. It announces all things without exception, just as a bell echoes
everywhere, even where it is unheard. This piece of paper, in all its
uniqueness and irreplaceability, requires *everything else* in order to exist.
Pulling out this piece of paper tugs at the fabric of all things without

exception. Thus, to paraphrase Nishitani (1982, 156), the fact that this paper is this paper is a fact in such a way as to involve at the same time the deliverance of all things in their original, Earthly countenance and interdependency (Sanskrit, *pratitya-samutpada*). This is why it makes a literally humiliating sort of sense to say that if this particular piece of paper did not exist, nothing would exist, for with the nonexistence of this paper, a constitutive element of all other things would be missing such that none of those things would *be* what they now are *with* the existence of this paper—without this piece of paper, *everything* would be changed. This is also why there is a peculiar disorientation involved in suddenly realizing that *any* object, even the most trivial of things, is in the center of this interdependency, with all things ordered around it. There is nothing special about this particular piece of paper *and* it is the absolute center of all things without exception. To push this disorientation further, *"the center is everywhere.* Each and every thing becomes the center of all things and, in that sense, becomes an absolute center. This is the absolute uniqueness of things, their reality" (Nishitani 1982, 146). And even further:

> That a thing *is*—its absolute autonomy—comes about only in unison with the subordination *of* all other things. It comes about only where the being of all other things, while remaining to the very end the being that it is, is emptied out. Moreover, this means that the autonomy of this one thing is only constituted through a subordination *to* all other things. Its autonomy comes about only on a standpoint from which it makes all other things to be what they are, and in so doing is emptied of its own being (148).

In short, and against the Scholastic notion of "substance," we have a simple, ecologically sane insight: "all things in the world are linked together, one way or the other. Not a single thing comes into being without some relationship to every other thing" (149).

Realizing this is not simply "knowing." It ushers the possibility of *our own* deliverance from self-presence, self-stimulation, and self-announcement, drawing our eyes downward to this precious Earth, silently announcing the *real* genealogy, the *real* generativity and interdependencies of things and of ourselves, not just the one that happens from the neck up with the aid of Husserl or Heidegger or Gadamer. Clearly, moving in this direction risks the possibility of a liberating, albeit painful humiliation of our unchecked desire to put our edifications and aspirations at *the* center of all things. We risk finding that it may be elevating of Hans-Georg Gadamer's *Warheit und Methode* to say that it has the integrity and fertility and unbetraying character of a handful of Earth.

III

This bell ringing in the empty sky is not interconnected with the entities that surround it through a filling of the intervening space with knickknacks, like some Victorian parlor: a peculiarly appropriate image for British Empiricism and the relentless proliferation of clutter from experimental designs, designed to fill the gap between one finding and the next through first methodically *isolating* some supposedly separate intervening variable —another knickknack found between two others—and then summoning up doubtful, dusty, mathematized interconnectedness of our own making between what we have first rendered isolated by the demands of the logic of empiricism. Thus the interconnectedness of things becomes like cobwebs in the dark. Life becomes sadly episodic, momentary in place and time, reduced to divisible minutia coupled with "a strange, almost occult yearning for the future" (Berry 1986, 56) when things will finally be okay. And, in inverse proportion to a diminishing sense of time and place, our sense of our own potency and importance becomes distended: as time becomes more and more televised-episodic, the most meagre remembrances become enormous and powerful in their resonances: as "personal space" holds sway as a sense of place, that things are all right can be confirmed by just looking around me—inside my house, inside my car, inside my head, everything is fine:

> The industrial conquistador, seated in his living room in the evening in front of the TV set, many miles from his work, can easily forget where he is and what he has done. He is everywhere and nowhere. Everything around him, everything on TV, tells him of his success: *his* comfort is the redemption of the world (Berry 1986, 52–53).
>
> In homes full of "conveniences" which signify that all is well, in an automated kitchen, in a gleaming, odourless bathroom, in year-round air-conditioning, in color TV, in an easy chair, the world is redeemed. If what God made can be made by humans into *this,* then what can be wrong (52)?

A bell muffled in a stuffy, enclosed space (or so the one *in* the enclosure can afford to believe).

IV

This bell ringing in the empty sky is interconnected with the entities around it; it can issue its resonance with all things, precisely because it is *not* crowded/surrounded/enclosed. The empty sky is not a vacuum to be filled. It is not *filled* with the sound of the bell. It doesn't *long* to be filled. It is not *deserted* or *alone* or *lonely*. It is empty in the sense that it is empty of self-existence (Sanskrit: *svabhava*): to be what it is, it empties out into all things without exception, it summons up an intimate, unenclosed indebtedness to the whole of the Earth.

This exquisite thing is part and apart, precisely the phrase used to elicit the paradoxical and irresolvable place of children in our lives (Smith 1988b, 175). Part and apart; not quite the same, but not simply different. Children are our kin, our kind, and we are drawn out to them by a natural affection (a "kindness") in the same way that the bell rings out while remaining itself; in the same way that the bell can ring out only *because* it remains itself. Only because of the empty sky can it have the room to ring out. Only because our resonant, generative, and ambiguous kinship is not an identity nor a difference can we be drawn to children with any natural affection. To sustain this natural affection, this kinship, requires that we neither desire children to be us, nor abandon them to their difference. Children are what they are because they are what they are not—their existence summons up us and the whole of the Earth as part of them, and apart from them. Akin.

Kin, kinship, kind, kindness, natural affection. And the parallel Sanskrit root is *gen*—generativity, regeneration, genesis, genealogy, generousness, that which is freely given. Affection, freely given. We should hesitate over this and imagine it as a way of calling attention to what grounds educational inquiry. Not isolatable substances, such that our work can speak of children as "that which needs nothing other than itself in order to exist," (which, with the advent of modern science, becomes "that which can be isolated as a variable that can be controlled, predicted, and manipulated such that its linear influences can be tracked"—note the relentless proliferation of stages and substages in developmental theories of child development, their manic multiplication a sign of the desire to be close to children, to bridge the difference, to embrace them). Rather, what grounds educational inquiry is affection, freely given; what grounds educational inquiry, what compels us to consider children *at all,* is that they are us, they are our kind, our kin, and we deeply desire to understand them and to understand ourselves in relation to them. Somehow, this makes madness of inquiry which *begins* with methodical isolation and severance and fragmentation of our lives and children's lives into severed competencies; somehow this accounts for the despair and pain that both we and children are experiencing in these times, perhaps especially in the area of education:

> Not only is fragmentation a disease, but the diseases of the disconnected parts are similar or analogous to one another. Thus, they memorialize their lost unity, their relation persisting in their disconnection. Any severance produces two wounds that are, among other things, the record of how the severed parts once fitted together (Berry 1986, 110–11).

Such fragmentation and isolation results in an image of inquiry as an out-of-control urban(e) sprawl which knows nothing of the moist Earthly ground it travels, upon which it relies, and which it covers over with a lifeless veneer of its own making. (I cannot help thinking of the clear and well-lit overhead projector images of models and flowcharts, each line well drawn and unequivocal, with all the difficult and generative resonances severed, it is hoped, once and for all.) Inquiry in education thus

> becomes more and more organized, but less and less orderly. No longer is human life [understood as] rising from the earth like a pyramid, broadly and considerately founded upon its sources. Now it scatters itself out in a reckless horizontal sprawl, like a disorderly city whose suburbs and pavements destroy the fields (Berry 1986, 21).

V

Understood as a living, Earthly relationship, this paradox of our lives with children—part and apart—is not struggling to resolve itself into well-drawn, unambiguous, unequivocal declarations. Rather, being mindful of this paradox and its generative irresolvability "makes meaningful and beautiful the primary paradoxes that human beings *have* to live with" (Snyder 1980, 29–30), the "original difficulty" (Caputo 1987) that resides in the issuing forth of new life in our midst. Children are a reminder of archaic debts, reminders of a *real* genealogy and the interlacing of that genealogy with all things. Perhaps this is why Gary Snyder maintains that we should live in a place as if we and our children, and their children, will live there a thousand years. This genealogy, this generativity, and this deep natural affection—the question of "kind"—does not and cannot simply string along a chromosomal thread, but exhales outward into the whole of the Earth, into a sense of place and space, coupled with an attunement to what is needed for a livable future. The whole Earth is our "kind," our kin. We are human, full of humus, and our natural affection cannot begin with isolating humanity as "that which needs nothing but itself in order to exist." Our humanity is not a substance. We are empty of self-existence and only as such—only interlaced with all things—can we be what we deeply are:

> With the interdependence of all things, cause and effect are no longer perceived as linear, but as a net, not a two-dimensional one, but a system of countless nets interwoven in all directions in a multidimensional space (Hahn 1988, 64).
>
> Even the very tiniest thing, to the extent that it "is," displays in its act of being the whole web of circuminsessional interpenetration that links all things together (Nishitani 1982, 149).

Whispering in this paradox of solitariness/exquisiteness and inter-relation is a sense of having a place which is not of my own making: it is granted a place, freely given, and mindfulness of this gift is, in a deep sense, a finding (Heidegger 1954, 1968). My being, even in my individuality, is an original blessing (Fox 1983), a gift—it is a *given,* and the most intimate and original response must be one of thanks for the naturally affectionate generosity of the Earth. It is mere hubris to consider ever finding ourselves out of debt, thinking a thought, living a life which owes nothing, which is perfectly self-enclosed. Such was the poverty and disaffection of Descartes's dream (Jardine 1990)—the unindebted thought thinking itself, self-present, out of relation to all things. True thinking does not and cannot repay this debt. True thinking simply heightens our sense of this original debt, this original gift, this original blessing. In receiving, a giving . . .

VI

A Bell Ringing in the Empty Sky: This image cannot live with the bizarre and convoluted horror pronounced so calmly, in the mid-nineteenth century by Arthur Schopenhauer:

> "The world is my representation": This is a truth valid with reference to every living and knowing being, although man alone can bring it into reflective, abstract consciousness. If he really does so, philosophical discernment has dawned on him. It then becomes clear and certain to him that he does not know a sun and an earth, but only an eye that sees a sun, a hand that feels an earth; that the world around him is there only as representation, in other words, only in reference to another thing, namely, that which represents, and this is himself (1966, 63).

At first glance, Schopenhauer's words might seem to foretell a sense of interdependency, but, in the end, it is only a dependency. To understand the sun, the Earth, requires only a reflective understanding of our powers to represent. Between sun and Earth there is an interdependency, but such interdependency is a matter of our representing, of our *bestowing* cosmos on chaos (as Jean Piaget so calmly defines the young child's actions on the world [1954, 1971b, 15]), such that interdependency is rooted in a deeper dependency on the one who represents. The Earth is ours to envisage, to make in our own image. Thus "the sickness of the West" begins. In our envisaging of the Earth, all we can see is this facade, and this facade gains its integrity by being linked back to human being—suddenly we, as humans, have no place on the Earth but have become that place in which all other things, the whole Earth, comes forth:

> Man sets himself up as the setting in which whatever is must henceforth set itself forth, must present itself. Man becomes the representative of that which is. What is decisive is that man expressly takes up this position as one constituted by himself . . . and that he makes it secure as the solid footing for a possible development of humanity. There begins that way of being human which mans the realm of human capacity as a domain given over to measuring and executing, for the purposes of gaining mastery of what is as a whole (Heidegger 1950, 1971b, 132).

Under the shadow of Western epistemology, "what is as a whole" possesses a singular center which dispenses the possibility of things and thus oversteps its own Earthly possibility with unearthly, transcendental zeal. The human being becomes the condition for the possibility of all things. The world becomes our representation as we solemnly become the singular representatives of all things. We become, here, the grand colonizers (it is no coincidence that Western epistemological heritage is European and that it is from a Eurocenter that colonization proceeded). We become the ones that savage those we consider unorganized, uncivilized, illogical, immoral, immature, by rendering them in our own image. We don't allow them a face (difference, here, must be fixed, for to be different is to fail to be at the center); we give them a facade of our own making. Deep in our Western heritage, and threading lines into contemporary educational theory and practice, there is a pleasant, attractive name for this colonization—we wish to understand.

VII

From Immanuel Kant's *Critique of Pure Reason* (1787, 1964), we have the fearful recoil from the decentering discoveries of Copernicus. Copernicus displaced the Earth as center and placed the sun at the center of the visible universe. The heavens no longer revolved around us. Kant's Copernican Revolution re-claimed what was lost in this movement:

> A light broke upon the students of nature. They learned that reason has insight only into that which it produces after a plan of its own, and that it must not allow itself to be kept, as it were, in nature's leading-strings, but must itself show the way with principles of judgement based on fixed laws, constraining nature to give answer to questions of reason's own determining. Reason . . . must approach nature in order to be taught by it. It must not, however, do so in the character of a pupil who listens to everything the teacher chooses to say, but of an appointed judge who compels the witnesses to answer questions which he had himself formulated. While reason must seek in nature, not fictitiously ascribe to it, whatever has to be learnt, if learnt at all, only from nature, it must adopt as its guide, in so seeking, that which it has itself put into nature (20).

To understand the essence of the Earth, one must understand the essence of Reason, for it is from Reason that the synthetic unity of nature-as-experienced, its interrelatedness, arises. To understand the Earth, we must seek what we have put there. Here, once again, is the deep hubris and deep faith of European Enlightenment—*only Reason is exquisite,* untouched, unsurrounded, ringing forth with no equal, graciously *bestowing* its forms on all things through its resonating synthetic acts.

VIII

The spirit of the Earth, its inner fire, its *logos,* has become ourselves. In this way, *eco logos* becomes *ego logos,* ecology becomes egology. Understanding ourselves need only refer to ourselves: to our skills and concepts, our mastery of requisite inner states of competence and performance, our skill at wielding methods and processes. We no longer need to rely on how our selves are reflected back to us from what surrounds us and houses us (*eco*), from what *things* have to say about *us.* The Earth has been silenced—we are the only resonance, the only and singular voice (see Habermas's [1972] characterization of scientific discourse as monologic).

Consider: *eco,* dwelling, abode. Consider Heidegger's definition:

> Dwelling is not primarily inhabiting but taking care of and creating that space within which something comes into its own and flourishes. Dwelling is primarily saving, in the older sense of setting something free to become itself, what it essentially is.. . . Dwelling is that which cares for things so that they essentially presence and come into their own (cited in Devall and Sessions 1985, 98–99).

And thus consider the legacy of Kant: the place where things are brought into their own and flourish is human consciousness and human articulation. Certainly, dwelling is that which cares for things, but since the integrity of things is a bestowal of Reason, caring for things, caring for this precious Earth becomes a matter of Reason caring for and heeding itself. The care involved in dwelling becomes meticulous and methodical carefulness. Dwelling becomes equated with the caution and suspicion of logicomathematical discourse. Dwelling becomes method. Or, consider the legacy of Nietzsche. Here, dwelling, the place where things are brought into their own, becomes self-declaration, self-announcement, desire and will. Lost is any sense of something being worth saying; gained is an ascendency of simply saying, the episodic scatter-shot of opinion and rapid-fire whole-language blabbering all at once. All at once, for it no longer matters if anyone really listens. Listeners are just raw meat, they just provide resistance and gutted nourishment which can increase my sense of power—I must consume the listener with my speech, demand attention, and all of this must be done quickly. I write this under the influence of practicum supervision, where many grade-one classrooms have become filled with a panicky rush where there is so very much being done and so little time for anything. There is no time to *dwell* over anything, to take care and consideration and deep attention to what you are doing (teacher or child), to prolong the sensuous richness of things, or the sonorous richness of words and melodies, to let the resonances ring out fully. There is no time for silence, for space, for seasonality, for these are merely lack of will, lack of diligence and effort, failures to produce.

IX

In case this seems to have veered too far from curricular and educational issues, in case this Kantian legacy seems lost on us in educational theory and practice, consider, from Jean Piaget's *Insights and Illusions of Philosophy* (1965, 1971a):

> Every relation between the living being and its environment has this particular characteristic: the former, instead of submitting passively to the latter, modifies it by imposing on it a certain structure of its own (118).
>
> One can feel very close to the spirit of Kantianism (and I believe I am close to it . . .). [However] the necessity characteristic of the syntheses becomes [in my work] a *terminus ad quem* and ceases to be as in [Kant] a *terminus a quo* (57).

In Piaget's work, the categories of Reason in Kant become the logico-mathematical operations that are the point *to* which (*terminus ad quem*) the development of knowledge tends, rather than the point *from* which (*terminus a quo*) knowledge proceeds. They are the operations that children are "destined to master" (Piaget 1952, 365)—a developmental destiny described by Piaget as a progressive "conquest of things" (363)—in their interactions with the world, for these operations are best adapted to the functional *processes* involved in that adaptation. Structures are thus "the products of a continuous activity which is immanent in them and of which they constitute the sequential moments of crystallization" (388). The peculiarity of the Kantian categories is that they constitute "an extension and perfection of all adaptive processes" (Piaget 1942, 1973, 7) insofar as they are perfectly adapted to the functional *a priori* of life itself, perfectly adapted, that is, to the "organizing activity inherent in life itself" (Piaget 1952, 19). In short, understanding becomes not so much a fixed set of structures or categories, but a way of operating on things; understanding becomes "construction"; the matters at hand become matters of *method* (Heidegger 1962, 1972). Logico-mathematical knowledge which, in Kant, described the *logos* of the phenomenal world, becomes in Piaget those set of operations or methods which best crystallize our self-regulating *proceedings "upon" the world* (proceedings which issue a "more or less rapid organization of sensorial data" (Piaget 1952, 369)—all that is given is scattered data with no integrity of their own; our proceedings cannot be violent for there is no integrity to violate). To understand the world is to operate on it and thereby constitute that world through such operations; to understand oneself is to understand how one operates. The triumph of the will over the Earth becomes developmentally sequenced. Reading the title of one of Piaget's works no longer arouses any qualms or suspicion or embarrassment or hesitation or humility—The Construction of Reality in the Child.

X

The universe is a dynamic fabric of interdependent events in which none is the fundamental entity. (Hahn 1988, 50).

But it is so difficult to actually *believe* this, to summon up the understanding of myself that allows this to come forward. I am not an *exception* to the interweaving, Earthly indebtedness and interdependency, but an instance of it. Believing this requires a deep sort of "letting go" that is not commonplace in Western aspirations. It entails a sort of simple, reverential understanding of things and our reliances on them.

But this also occasions a realization of the fact that "the self in its original countenance" (Nishitani 1982, 91) is an Earthly self, full of the humus that interweaves with all things:

> The self is here at the home-ground of all things. It is itself a home-ground where
> everything becomes manifest as what it is, where all things are assembled together
> into a "world." this must be a standpoint where one sees one's own self in all
> things, in living things, in hills and rivers and towns and hamlets, tiles and stones,
> and loves these things "as oneself" (Nishitani 1982, 280–81).

There is a peculiar turn, here, that distinguishes this passage from those of
Eurocentrism. Seeing oneself in all things is not a matter of the violent
colonization of all things, as if to see oneself "in" all things is to have
impregnated them, to have pissed on every tree, surveyed every acre. It is not
a matter of enslaving all things and making them indebted to me. It is, rather,
a matter of recognizing my indebtedness to *them*. Loving all things as
oneself is a matter of *emptying out* that self, of *giving away* one's
center(ed)ness. This means, at first twist, becoming "nothing." But it also
means, at once, becoming all things:

> I gave up trying to carry on an intellectual interior life separate from the work, and
> I said the hell with it, I'll just work. And instead of losing something, I got
> something much greater. By just working, I found myself being completely there,
> having the whole mountain inside of me, and finally having a whole language
> inside of me that became one with the rocks and with the trees (Snyder 1980, 8).

This is so difficult to realize, even gazing out the window to these trees, this
snow, that all these things are not there *for me*. This is the sting of ecological
insight. But in this insight is contained something precious. Once I deeply
realize that all these things are not here for me, they begin, so to speak, to
turn away; they no longer pay any special attention to me; they are not
formed up and tarted up for my perusal. Suddenly, I am no longer the
displaced, post-Copernican, questioning stranger, the colonizing interloper,
the condescending intruder to whom attention must be paid and who can
make demands without attention to where I actually am, without attention
to what is *already at work without me*. Suddenly, I *belong* here. I *live* here.
This Earth is my *home*. I can finally experience my being as resonant,
indebted, and interwoven with these things. I am no longer the judge posing
questions and demanding silence. I can begin, as ecology suggests, to
become deeply conversant with things, listening, asking, responding,
inhaling, and exhaling. Such conversation is only possible insofar as it issues

up out of a sense of the integrity of my interlocutor—the Earth in all its wondrous articulations.

XI

There are afoot forms of philosophical/educational discourse which speak on behalf of deconstructing the violent, retentive centrations that have ruled our lives—logocentrism, phallocentrism, Eurocentrism, anthropocentrism, adultocentrism. And often, such language is disorienting, intentionally disruptive, disrespectful, unsettling, profane, sarcastic, destructive, negative.

Question: What is such work *for?* The danger we face is that we may come to believe, with the "postmodernist" disassembling of violent centrations, that *nothing* is the center, that all things are trivial and episodic and unconnected, and that only a vague, exhausted, and, in its own way, violent licentiousness is possible.

A bell ringing in the empty sky: all things are at the center. All things are exquisite. What hopefully emerges from such disassembling of the clangor are those archaic sounds and resonances that were not of our own assembly in the first place. What hopefully emerges out from under the exhaustion is a deep love and compassion for all things.

XII

The real work is what we really do. And what our lives really are. And if we can live the work we have to do, knowing that we are real, and that the world is real, then it becomes right. And that's the real work: to make the world as real as it is and to find ourselves as real as we are within it (Snyder 1980, 82).

How can our eyes be drawn back to Earth and to the real work of attending and attuning to the *humus* out of which our humanity has arisen? We cannot confront this epistemological legacy head on, for it already defines its foes. But neither will it do to fantasize about wide-open spaces or escapes to elsewhere. For there is no "elsewhere." It is this place, this space that are at issue in drawing our eyes back to Earth. This is the juncture at which the real work becomes an economic and political matter. It is at this juncture that the real work becomes physically, spiritually, linguistically, culturally, and autobiographically bound in ways that no heavenward glance can surmount. The *real* critique does not fall from above; it issues up out of the Earth we actually walk, the life we actually *find* ourselves living. This school, this classroom, this moment with this child—these are the resonant moments that are full of our indebtedness and fully worthy of our attention. If a bell ringing in the empty sky becomes a woozy vision of Japanese exotica, the real work is lost and the talk is mere pornography and titillation with novelty items. Although the problem is "where do you put your feet down; where do you raise your children; what do you do with your hands?" (Snyder 1980, 66), the problem is also how we find ourselves enabled or disabled to respond to, to even *raise* these questions. The real work is right here.

So it will not do to deny that this paper is written by a male of European/Caucasian descent, someone economically well-off, living on an acreage far outside a city, with plenty of trees and sun and sky, with a reasonably secure job that has built-in luxuries, built-in time and space for mindfulness and Earthly tempos. I am one for whom these words are easy to pronounce, for they do not have to clamor up out of the deadly white-noise silencings which define and defile so many. "How can I, as an educator, fulfill my responsibility to my own people; my own people whom I love yet who, like I do, live under an economic and epistemological dispensation which is the *problem* for most of the world" (Smith, 1988d, 92)?

I hope my son can live a life which is *not* just a niche in the clutter. This hope is difficult to maintain when I see how cluttered his school life is *already* and when I hear the ecological rumblings in the distance. I hope he finds the empty space required for the natural affections and kinships of speech and experience and understanding to come forth. I hope he becomes deeply conversant with this precious Earth. I hope someday that he will understand that he, in his good fortune, is indebted to all things without exception. I'm afraid that he'll *have* to see someday that his good fortune is the *problem* for most of the world and that some of those debts are coming due.

And I am frightened that I won't know what to say when he asks what I have actually *done* to live in the spirit of a loving and compassionate indebtedness to all things; what he will say if he discovers that it is some of *my* ignored debts that he must repay, perhaps with his life (after all, where exactly *is* the dioxin produced by the bleaching of these very pages?). *This* is the real work of my life as an educator, no matter what I say.

A bell ringing in the empty sky.

CHAPTER SIX

Immanuel Kant, Jean Piaget, and the
Rage for Order: Ecological Hints of the
Colonial Spirit in Pedagogy (1994)

Introduction

In these ecologically desperate times, we are confronted with the painful task of recollecting those images of ourselves and our Earth that we have come to live with in our educational theories and practices. We are being forced by the sometimes gentle, sometimes violent guidances of the Earth, to reconsider what we understand ourselves to be and what, therefore, we wish our children to become. We are being forced back into questions that form the core of pedagogy itself, and pedagogy, too, is being slipped back into its element.

Ecology is reminding us that however aspiring our pedagogical theories might be, if they speak implicitly against or in spite of the fleshy conditions under which we may draw real breath to actually speak such aspirations *at all,* such theories may be logically and epistemologically coherent, but may contain that bewildering, low-level dis-ease that we have come to understand as modern life. Such pedagogies may not be precisely *false* according to some norm of logical consistency, empirical verifiability, or the like. They may, however, betray a sort of insanity that Wendell Berry (1986) hints at so eloquently:

> It is impossible to divorce the question of what we do from the question of where we are—or, rather, where we think we are. That no sane creature befouls its own

nest is accepted as generally true. What we conceive to be our nest, where we think it is, are therefore questions of the greatest importance (51).

This paper explores some of the threads that influence what we consider our place to be in the area of pedagogical theory. What follows is a discussion of the work of Immanuel Kant, followed by a parallel explication of the Kantianism inherent in the work of Jean Piaget. In both cases we are presented with a bloated image of the importance of humanity in giving the Earth integrity and meaning.

Although this may anticipate too much of what follows, we cannot deny, along with Piaget, that we structure our experiences in light of the categories and concepts ("schemata") at our disposal and therefore "make" that experience meaningful. However, in the face of this realization, it is not enough to blithely wallow in the constructions we have made, for they *also* describe the limits and dangers we face. We tend to impose constructs upon what comes to meet us, giving it order(s). We tend, too easily, to allow our understanding to be the right of passage (*colon*) for all things, forgetting the dependencies that draw our lives back into the Earth, forgetting that we, as living beings full of humus, pass through the Earth's flesh and are "ordered" by it.

Jean Piaget is a rather obvious choice for critical consideration in the area of pedagogy. His work has come to infuse educational theory and practice in a profound way, and his legacy persists in spite of the quarrels that have ensued in the literature as to the truth or falsity of this or that aspect of Piagetian theory. Those quarrels will not be taken up here. They are for the friends and enemies of Piaget to take up in ways appropriate to the "inevitable advance" (Piaget 1965, 1971a, 102) of empirical work. Here, our concern is with the lingering images of Piagetianism that we have come to live with in the sphere of pedagogy.

The choice of Immanuel Kant is, on the surface, a less obvious one. One reason for this choice is that Kant's *Critique of Pure Reason* (1787, 1964) marks a revolution in our understanding of ourselves and our place on Earth, the consequences of which we are still living out. Instead of the Earth putting us into question with the mysteries it poses about our dependencies, our frailties and our kinships with all life, it is we who put questions to nature. The ambiguous dependencies of the flesh become simply another object in the inert objective mechanisms of "nature." Reason is conceived as *a priori,* as independent of Earthly inheritance. And our belief that we might somehow live independently of our Earthly inheritance is a perfect segue into understanding the ecological rumblings we are now experiencing.

Another reason for this choice of Immanuel Kant is Piaget's explicit naming of his own work as standing in the Kantian tradition (1965, 1971a, 57). In Piaget's work, we may recognize the fleshy inheritance that logico-mathematical knowledge owes to the neonate and the embodied knowings of sensory-motor knowledge. But the progress of development is, in effect, the sequentially shedding of our reliances upon such an inheritance: "climbing up into our heads" (Le Guin 1987, 11), and, more pointedly, teaching our children to do the same. In fact, Piaget names our children as the ones who are *destined* to do the same (Piaget 1952, 372).

The explorations which follow are not intended to do justice to the *intent* of the works considered. They are meant to bring out an ecologically disastrous undercurrent that haunts these works and haunts the lives we are living out with our children. It is far too late to expect satisfaction from blaming authors in the past for not foreseeing current concerns. What we can expect from considering their works interpretively is a way of opening up and reading the signs we now confront. We are living out a logic that is centuries old (Berman 1984, 8) and the signs we now confront must be read back into this logic.

Following these readings of Kant and Piaget, consideration will be given to Wendell Berry's (1983) distinction between "chaos" and "mystery" as images of that which falls outside of the boundaries of the known, and how the notion of mystery, far from being an esoteric notion, leads to a sense of practical propriety, a sense of living wisely and delicately on the Earth, mindful to the patterns that pertain whether we know it or not.

Immanuel Kant: Breaking Free of Nature's "Leading Strings"

Immanuel Kant's so-called "Copernican Revolution" (1787, 1964, 22) in philosophy reclaimed at the *epistemological* level what was lost in the *cosmological* level in the work of Copernicus. Copernicus displaced the Earth as the center of Creation and placed the sun at the center of the visible universe, thus decentering humanity from its special cosmological place.

Kant reclaimed the center by putting human Reason at the center of the *knowable* Universe: anything knowable refers back to the conditions of knowability which are determined by the essential character of Reason. But there is an extra step to watch for in this reclaiming. Human Reason, as a synthesizing faculty, bestows order upon the Universe, turning the chaotic influx of experience into a knowable (i.e., orderly) *cosmos.* Knowledge becomes conceived as *constructive of a world.* This additional step profoundly shifts the logic of the Cartesian legacy in which Kant's work stands.

Descartes (c.1640, 1955) rendered the subject *worldless* through his methodical doubt (90) and the world was retrieved only through God's benevolent guaranteeing of the objective reality of the clear and distinct ideas that this worldless subject thinks (107–26). The subject, in Kant's work, is not *worldless.* But neither has it retrieved the world itself through Divine intervention. Rather, *the human subject lives in a world of its own making.* Given this, the human subject can do with the world what it wishes, since the world as experienced is of its own "doing" in the first place.

Consider, from the preface to Immanuel Kant's *Critique of Pure Reason* (1787, 1964):

> A light broke upon the students of nature. They learned that reason has insight only into that which it produces after a plan of its own, and that it must not allow itself to be kept, as it were, in nature's leading-strings, but must itself show the way with principles of judgement based on fixed laws, constraining nature to give answer to questions of reason's own determining. Reason . . . must approach nature in order to be taught by it. It must not, however, do so in the character of a pupil who listens to everything the teacher chooses to say, but of an appointed judge who compels the witnesses to answer questions which he had himself formulated. While reason must seek in nature, not fictitiously ascribe to it, whatever has to be learnt, if learnt at all, only from nature, it must adopt as its guide, in so seeking, that which it has itself put into nature (20).

There are two echoes in this passage that are enticing. Both contain implicit images of pedagogy. One is a clear echo of the Enlightenment call that Kant himself (1784, 1983) expressed:

> *Enlightenment is man's emergence from his self-imposed immaturity. Immaturity* is the inability to use one's understanding without guidance from another. This immaturity is *self-imposed* when its cause lies not in a lack of understanding, but

in a lack of resolve and courage to use it without guidance from another. *Sapere Aude!:* "Have courage to use your own understanding!"—that is the motto of enlightenment (41).

Kant's Copernican Revolution elevated human Reason to the point of being its own guide, of needing no guidance from another. We therefore find hints in the above-cited passage from the *Critique of Pure Reason* of the pupil turning away from his teachers and their guidance (one meaning of "leading strings" is "pupilage" or being lead or taught by another). Under the Enlightenment ideal, *any* guidance is ruled out as indicative of childishness or immaturity. We must approach the Earth unindebted, "self-guiding."

Another image buried here is one of the boy cutting the leading-strings (or "apron strings") that bind him to his mother—an implicit denial of any lingering dependency on the one who gives us birth. If we admit that the Earth gives us birth, then Reason remains somehow indebted to another. This again despoils the Enlightenment ideal of a humanity guided by Reason alone, somehow independent of its animate but irrational humus ("leading-strings" names a cord used to lead animals). Once the leading strings are cut, it is we now who take the lead, who can "stand on our own two feet" ("leading-strings" were used to teach children to stand and walk; it is also used as a metaphor for "dependency"). As with Descartes's methodical doubt, this metaphor of Kant's embodies the denial that the Earth somehow guides us, sustains us, and bears us up. Kant's work stands, to this extent, in the shadow of the Cartesian ideal of a subject who is severed and separate from the Earth. Now separate and alien, the Earth no longer houses us but stands before us as an object. This object can henceforth be subjected to the subject's logico-mathematical demands for clarity and distinctness from all that comes to meet it. However:

> If a kind of Cartesian ideal were ever completely fulfilled (i.e., if the whole of nature were only what can be explained in terms of mathematical relationships) then we would look at the world with that fearful sense of alienation, with that utter loss of reality with which a future schizophrenic child looks at his mother. A machine cannot give birth (Bordo 1987, 97).

We need to consider in more detail how this schizophrenia is possible, for it has profound effects on pedagogy. As Susan Bordo (1987) goes on to suggest:

We are all familiar with the dominant . . . themes of starting anew, alone, without influence from the past or other people, with the guidance of reason alone. The specific origins of obscurity in our thinking are the appetites, the influence of our teachers, and the "prejudices" of childhood. The purification of the relation between knower and known requires the repudiation of childhood . . . through a deliberate and methodical reversal of all the prejudices acquired within it, and a beginning anew with reason as one's only parent (97–98).

Should threads of this legacy infuse educational theory and practice, we are left with a bizarre prospect. It may be that a great deal of work in educational inquiry begins with an implicit, unvoiced repudiation of childhood. Such a repudiation might make it possible to render children into a controllable, predictable and manageable *object,* and such rendering certainly has a place in our living relationships with children. But it at once rends the tenuous, ambiguous threads that make children our "kin," our "kind," and that make us and our children *of* the Earth. In striving to break these threads in the name of a Reason able to act without guidance or hesitation or any sense of "kind-ness" towards what comes to meet it, ecological disaster is already foretold.

Steps in the Breaking of the Strings:
The *"A Priori"* as the "Unborn(e)"

Against empiricism, Immanuel Kant maintained that mathematics, Euclidean geometry, and formal logic cannot be derived from experience, since all that experience can allow us to derive from it are empirical generalizations which can never attain the universality and necessity requisite of logic and mathematics. For empiricism, each distinguishable empirical moment is separate and distinct. "Experience" is a chaotic rush of instances without form or figure. For experience to have any shape at all requires our concerted action *upon* it.

For empiricism, the only warrantable action upon experience is *posterior* to it, *a posteriori.* The only warrantable action upon experience is a generalization *from* the instances of experience. Such generalizations are always subject to the continuing influx of experiences. The next instance can prove incorrect the empirical generalization thus far derived from past experiences. They are therefore not universal but general(ized). They are not necessary but contingent and dependent upon the next instance.

Kant's work began elsewhere. He began by affirming the *de facto* existence of logic and mathematics as forms of knowledge which are universal and necessary and therefore cannot be derived from generalizations from empirical experience. Such generalizations will not yield the necessity and universality requisite of logic and mathematics. Kant maintained that the de facto universality and necessity of logic and mathematics was an indication that they were not *a posteriori* but *a priori*—they were not *derived from* experience, but rather were *prior conditions of the essential operations of thought itself,* conditions henceforth demanded of experience by the nature of Reason itself. These conditions shape the "questions of Reason's own determining."

Formal logic delineates the essential, *a priori* (i.e., universal and necessary) interrelations between one thought and another. It describes the organization of thought itself (see Piaget 1952, 15). Such a formal logic becomes a transcendental logic (i.e., a relation, not simply between one thought and another, but between thought and things; i.e., a logic whereby the subject transcends itself; i.e., the logic inherent in how the subject can be presented with an object in light of the subject's own epistemic conditions; i.e., the logic of representation) when applied to the influx of experience.

Thus, for example, in formal logic, we have the hypothetical judgement "If A, then B" (Kant 1787, 1964, 107 ff.), which describes one of the ways that one thought can be related to another—hypothetically. Another formal structure of Reason is the predicative or attributive form of judgement "A is B" (107 ff.). With such formal-logical judgements, we are still dealing with the relation between one thought and another. We are still getting a sort of "filling out" of the contours of Descartes's "I think"—this is *how* "I think," (i.e., the essential forms of how one thought fits with another with clarity and distinctness). These essential relations of one thought to another are what is called the Table of Judgments in Kant's *Critique of Pure Reason* (107).

Formal logic becomes a transcendental logic when we impose the forms of judgment upon sensory experience and "en-form" it. Simply put, when we think *about* our (empirical) experience, *how* we think makes en-forming demands on that experience, giving it "form" (or, to use Piaget's term, "schematizing" it). Parallel to the Table of Judgments, then, we have *those very same forms* now in their en-forming, transcendental function: this is the Table of Categories in Kant's *Critique of Pure Reason* (113 ff.). The hypothetical judgment "If A then B" becomes, in its transcendental function, the category of causality. And since the hypothetical form is essential to the nature of our thinking, it becomes essential to the nature of any possible object *about which* we are thinking: "All events have a cause." For example, the sound at the door *must* have been caused by something, even if I don't know precisely *what* the cause is. That is, *prior* to searching out the empirical information as to what caused the knock on the door, I know, *a priori, that* it had a cause, and this *a priori* knowing is what guides the empirical search for *what* the cause might be.

Transcendental logic thus describes the synthetic relation between objects of experience, prescribing the logic of objectivity. We must revert to empirical experience to find out, for example, precisely *what* the cause is of a particular event ("the most the understanding can achieve *a priori* is to anticipate the form of a possible experience in general" [Kant 1787, 1964, 264]), but *that it has a cause* is a necessary and universal condition of knowledge, and therefore is a necessary and universal condition of any possible object of knowledge.

In this way, it is no longer especially peculiar to say that "the *a priori* conditions of a possible experience in general are at the same time conditions of the possibility of objects of experience" (Kant 1787, 1964, 138). We end up with Reason remaining within the parameters of its own universal and necessary conditions and then "imposing" these conditions upon the influx of experience. Whatever comes to meet us, we *already know* what its essential structures and forms will be, because that meeting is enformative before and as a condition of it being in-formative. That is to say, without the universal and necessary (i.e., *a priori*) en-forming knowledge *that* the sound at the door has a cause, collecting information about *what* the cause is would not be possible. This is the same as in Piagetian theory: once the child develops the schema of "number," for example, then this particular number of jelly beans can be known. Once experience is en-formed with the *possibility* of number, information about the *actual* number of this and that can be gleaned. We must learn from nature what can be learned only from it (what caused the sound at the door; how many jelly beans there are), but we must take as our guide what we have put there ourselves (*that* events have a cause; *that* things are enumerable). Here begins the frightening set of turns the legacy of which we are living out:

> . . . the order and regularity in [what] we call *nature,* we ourselves introduce. We could never find [such orderliness and regularity] . . . had not we ourselves, or the nature of our mind, originally set them there (147).
>
> [Human] understanding is itself the lawgiver of nature. Save through it, nature, that is, synthetic unity of the manifold of appearances according to rules [imposed by Reason itself] should not exist at all (148).

In the Kantian legacy, we give cosmos to the chaos of experience. We *demand* orderliness and regularity and, not finding it there before us in what comes to meet us, we *give* (it) order(s).

But here is the final twist that puts Kant's work beyond the Cartesianism. Descartes longed to have some guarantee that the order(s) he found so clear and distinct were really the order(s) of things themselves, things without us. Through the revealing of the systematically en-forming nature of Reason, Kant turns away from this longing. Consider:

> That nature should direct itself [in] conformity to law, sounds very strange and absurd. But . . . consider that this nature is not a thing in itself but is merely an aggregate of appearances, so many representations of the mind (140).
>
> The question arises . . . how it can be conceivable that nature should have to proceed in accordance with categories which . . . are not derived from it, and do not mold themselves on its pattern? The solution of this seeming enigma is as follows. Things in themselves would necessarily, apart from any understanding that knows them, conform to laws of their own. But appearances are only representations of things that are unknown as regards what they may be in themselves. As mere representations, they are subject to no law of connection save that which the connecting faculty prescribes (178).

Things that are in relation to us ("objects of experience") are not things in themselves. To *know* something is to *put* it into relation with us. Things as known are not things in themselves. We can never know things in themselves, those things that might "conform to laws of their own." We can only know that which Reason produces after a plan of its own. We live, it seems, in a world of our own making that has ourselves as the center. With this last move, the phenomenal world (i.e., the world en-formed by Reason) becomes self-enclosed and self-referential. "Accordingly, the spontaneity of understanding becomes the formative principle of receptive matter, and in one stroke we have the old mythology of an intellect which glues and rigs together the world's matter with its own forms" (Heidegger 1925, 1985, 70).

Nature "in itself" may be, in some strangely meaningless way, preserved from the hubris of Reason by being rendered "unknowable." But as a correlate to such preservation, the phenomenal world becomes a closed system which has Reason as its master, as that to which the system is answerable. Reason becomes answerable only to itself. It becomes, to use Piagetian terminology, "self-regulating."

Through the twists and turns of the Kantian legacy, the spirit of the Earth, its inner fire, its *logos,* becomes ourselves. Eco-logos has become ego-logos. Consider: *eco,* dwelling, abode. Consider Heidegger's definition:

> Dwelling is not primarily inhabiting but taking care of and creating a space within which something comes into its own and flourishes. Dwelling is primarily saving, in the older sense of setting something free to become itself, what it essentially is. . . . Dwelling is that which cares for things so that they essentially presence and come into their own (cited in Devall and Sessions 1985, 98–99).

In the Kantian legacy, we become the ones to whom the Earth must present itself. Reason creates and controls and delimits the "space" within which things are brought into their own and flourish. Things come into their own only within the orbit of human dominion. Kant's work thus confirmed at an epistemological level what had been a long-standing belief in the Christian, European culture of his day: the Earth is there *for us.*

It is at this juncture that the metaphor of colonialism becomes potent. (European) humanity, conceived as cleaving to Reason, becomes the home (office) from which orders are given. What comes to meet us is colonized with demands for unequivocal clarity and distinctness. Reason becomes the right of passage (*colon*—the home office is where the colonized must get a "pass") for the Earth. Reason decides what will "pass" and what will not and therefore, since the Earth gains integrity from us, we can colonize it without hesitation, for this simply fulfills what is already prescribed *a priori.* It is not a coincidence that this epistemological inheritance is European and that it is from Europe that colonization proceeded. It is no coincidence that we might, on the epistemological level, believe that we are the ones that can warrantably savage those we consider undeveloped, disorderly, unorganized, unruly, unmethodical, immature, unreliable, unreasonable, possessing no indigenous integrity, in need of our gracious bestowal of order, needing to be "whipped into shape," since this belief regarding epistemology is tellingly synchronous with the cultural belief in the warrantability of cultural colonization of the savage lands of Africa and North America. In fact:

> In his 1772 lectures on philosophical anthropology at the University of Konigsberg, Kant proclaimed that the American Indians "are incapable of civilization." He described them as having "no motive force, for they are without affection and passion. They are not drawn to one another by love, and are thus unfruitful. They hardly speak at all, never caress one another, care about nothing, and are lazy." In a note in his lecture he foreshadowed two long centuries of racist thought in Germany when we wrote that the Indians "are incapable of governing themselves" and are "destined for extermination" (Weatherford 1988, 127).

Tighten this down one more turn and we find precisely these images operative in what Alice Miller (1989) called the "black pedagogies" of the seventeenth and eighteenth centuries: children are the wild and willful ones who must be made to "mind," who must be "taught a lesson." Another turn and we have the disturbing work of Ashis Nandy (1987), who documents how the colonized have been understood as children in need of guidance. And the graciousness of such colonization is always and everywhere done under the most noble of rubrics, so gently understated in the title of Alice Miller's (1989) book: *For Your Own Good.*

Jean Piaget and the Bearing of Reason

Jean Piaget's genetic epistemology can be read as working exactly against this violent pedagogic/epistemic legacy. It is his work that systematically uncovered the fact that children's understandings of the world go through identifiable stages, each of which has *its own* integrity and viability and each of which must be understood in relation to its indigenous organization. But then, too, Piaget maintains that these identifiable stages in children's thinking are, so to speak, "unstable." They are destined to be "destabilized" by the course of experience and sequentially restabilized at a more inclusive and stable level of development. In short, children are destined, in the long run, to become able to participate in the workings of logico-mathematical knowledge. Piaget simply demonstrates that this destiny cannot be worked out through violent exigenous action and colonization. Development has its own course and its own time.

But Piaget's genetic epistemology also stands, in part, in the Kantian tradition, and issues of colonialism erupt in it in a roundabout way. Consider, from Jean Piaget's *Insights and Illusions of Philosophy* (1965, 1971a):

> One can feel very close to the spirit of Kantianism (and I believe I am close to it. . . .) [However] the necessity characteristic of the syntheses [of Reason] becomes [in my work] a *terminus ad quem* and ceases to be as in [Kant] a *terminus a quo* (57).

Or, from his *Origins of Intelligence in Children* (1952):

> Every relation between the living being and its environment has this particular characteristic: the former, instead of submitting passively to the latter, modifies it by imposing on it a certain structure of its own (118).

In Piaget's work, the categories of Reason in Kant become the logico-mathematical operations that are the point to which (*terminus ad quem*) the development of knowledge tends, rather than the point from which (*terminus a quo*) knowledge proceeds. They are the operations that children are "destined to master" (Piaget 1952, 365) in the development of their interactions with the world—a development tellingly described by Piaget as a progressive "conquest of things" (363).

For Piaget, every interaction between the human organism and the world is an active organization of the world by the organism. That is to say, we impose upon what comes to meet us certain schematizations, "giving" our experience "order." However, unlike Kant, Piaget does not contend that the structures of logico-mathematical knowledge which describe the functioning of objective science are *a apriori* to any possible experience. Rather, they are *a priori* to the experience of any fully developed *adult* operating at the developmental level of logico-mathematical operations. In short, the structures that Kant took to be universal and necessary, Piaget demonstrates to have a sequential development in the life of the child. But if these structures or categories are shown by Piaget to be a *terminus ad quem,* not a *terminus a quo* of development, it seems, at first glance, that they cannot be universal and necessary and that Piaget's work might fall prey to the sort of critique of empiricism that Kantianism provides.

Piaget gets around this quandary in an interesting way. What is *a priori* in Piaget's work is not a set of structures or categories, but rather a set of functions—assimilation, accommodation, and equilibration. The structures evident in the life of the developing child are thus not fixed and unchanging from birth. Rather, they are "the products of a continuous activity which is immanent in them and of which they constitute the sequential moments of crystallization" (1952, 388). The developmental sequence of structures in Piaget's work is a sequence of plateaus of equilibrium, plateaus of stability in the organization of the functions of assimilation and accommodation. One set of structures replaces another because the latter set is more stable and inclusive, more *equilibrated,* than the former.

It is in this way that the invariant, functional *a priori* remains continuous (in that sense, universal and necessary) throughout development and throughout the changes of structure or developmental level. But more than this, this functional *a priori* of assimilation, accommodation, and equilibration gives directionality to those changes of structure, making them more than a series of random changes. Children don't simply *change:* they *develop.* And it is this functional *a priori* which

> orients the whole of the successive structures which the mind will then work out in contact with reality. It will thus play the role that [Kant] assigned to the *a priori:* that is to say, it [the *functional a priori*] will impose on the structures certain necessary and irreducible conditions. Only the mistake has sometimes been made of regarding the *a priori* as consisting in structures existing ready-made from the beginning of development, whereas if the functional invariant of thought is at work in the most primitive stages, it is only little by little that it impresses itself on consciousness due to the elaboration of structures which are increasingly adapted *to the function itself* (Piaget 1952, 3).

Logico-mathematical knowledge which, in Kant, described the *a priori* structures or categories of Reason that are imposed upon the influx of experience, in Piaget describes the purest expression or crystallization of an equilibrated set of *a priori* functions. At each stage of development, we imposed the schemata or structures we possess upon our experience and organize it accordingly—as Piaget (1954, 1971b) has put it, we impose cosmos on the chaos of experience (xii). Over the course of this influx of experience, elements of that experience will not be able to be schematized by the structures thus far developed and will cause "dis-equilibration" in those structures. In other words, some incoming experience will not be able to be assimilated into already-existing structures, nor will it be able to be accommodated to by already existing structures. Such dis-equilibration leads to an adaptive recrystallization of the functions of assimilation and accommodation into a more inclusive and more stable organization/ structure.

At first glance, it might appear that the sequence of developmentally emerging structures is oriented towards better and better adaptation to the Earth, thus making Piaget's work pointedly ecological at its core. In fact, this is not the case. If we carefully reread the extended passage cited above, we find that the sequential development of structures is better and better adapted, not to the Earth, but to *the inevitabilities of adaptation itself.* That is to say, it is the functions of assimilation, accommodation, and equilibration that are *a priori,* and the best adapted structures (most stable and most inclusive) are the ones that are best adapted "to the functioning itself." Development is oriented, therefore, towards better and better adaptation to the inevitable "organizing activity inherent in life itself" (Piaget 1952, 19). The peculiarity of the Kantian categories is that they constitute "an extension and perfection of all adaptive processes" (Piaget 1942, 1973, 7) insofar as they are perfectly adapted to this inevitable organizing activity. In this way, the Kantian categories take on the appearance of universality and necessity at the end of development because they are perfect expressions of that which *is* universal and necessary: the *functions* of assimilation, accommodation, and equilibration.

This helps explain why the sequence of development moves, in Piaget's work, from sensori-motor knowledge, to preoperational to concrete operational to formal-logical operational knowledge. At the highest level of development we have logico-mathematical knowledge which is, in essence, knowledge of the operation of knowledge itself, knowledge, that is, of the *functioning* that has been going on all along. At the level of logic and mathematics, however, we "proceed by the application of perfectly explicit rules, these rules being, of course, the very ones that define the structure under consideration" (Piaget 1968, 1970, 15). That is to say, at the level of logic and mathematics, the rules for *doing* the operations of logic and mathematics are precisely the rules *upon which* one operates. Logic and mathematics are thus perfectly equilibrated, for there is no longer any difference between the operator (understood as an anonymous epistemic subject), the operations, and the operand. The method of operation and that upon which we are operating is the same: the forms of logico-mathematical thinking are the very forms we are thinking *about* when doing logic or mathematics. Given this lovely self-enclosure, it is no wonder that logic and mathematics take on the appearance of being unborne, for they no longer owe debts to anything outside of the orbit of their operations. Development, in Piaget's work, becomes conceivable as the sequential shedding of such debts, which, once fully achieved, can act unindebted.

Logic and mathematics are now conceived, in Piaget's work, as organizations of operations rather than as fixed structures. Put another way, it is *scientific method,* operating as it does within the bounds of logic and mathematics, which poses questions of its own determining and demands of what comes to meet it that it give answer. It is the *functioning* of this method that is now *a priori.*

Jean Piaget's work thus leaves us with a paradox: children do not think like us, but to understand these ones who are other than us, we must proceed methodically and render this difference into a controllable, predictable, and manageable object of scientific discourse. At issue is not simply whether Piaget's work constitutes a form of colonization of children through the use of logico-mathematical knowledge as both a model of development and as a method for understanding that development. The issue is also whether Piaget's work, standing as it does within the Kantian legacy, constitutes an unintended apologetic for raising our children to *become* colonizers—ones destined to believe that they give the Earth meaning through their activities.

Concluding Remarks: Recovering a Sense of Propriety

Ecological awareness begins and remains fast within a paradox regarding human life. We can do the impossible:

> The unnoticeable law of the Earth preserves the Earth in the sufficiency of the emerging and perishing of all things in the allotted sphere of the possible which everything follows and yet nothing knows. The birch tree never oversteps its possibility. It is [human will] which drives the Earth beyond the sphere of its possibility into things that are no longer a possibility and are thus the impossible. It is one thing to just use the Earth, another to receive the blessing of the Earth and to become at home in the law of this reception in order to shepherd the mystery and watch over the inviolability of the possible (Heidegger 1936–46, 1987, 109).

"The inviolability of the possible" here is not commensurate with what we can do, assuming "that the human prerogative is unlimited, that we must do whatever we have the power to do. What is lacking [in such an assumption] is the idea that humans have a place, and that this place is limited by responsibility on the one hand and by humility on the other" (Berry 1983, 54–55). Human action, which both Kant and Piaget, each in different ways, put at the center of their considerations, can, so to speak, spiral out of order, out of proportion, breaking the delicate threads of kinship that might make our actions possible in the first place. We can do the impossible. We can, for example, destroy the ability of the Earth to produce the very oxygen we need in order to pursue such destruction.

Thus, ecology concerns not what we can do (in some utopian, Enlightenment-ideal sense, which is literally no-place), but what is proper, what is properly responsive to the place in which we find ourselves, those actions which have a sense of propriety, those actions which are "fitting," and which issue up out of a place as a considerate response to that place (i.e., a response which somehow acts in accordance with the sustainability of that response).

But then how are we to know what responses and actions have propriety? In a wonderful essay entitled "A Letter to Wes Jackson (1987)," Wendell Berry speaks of the urgent need to distinguish between the notion of mystery and the notion of chaos. The notion of chaos as a name for what goes beyond the sphere of what is known assumes that the sphere of knowledge is a limited one and that the existence of order is co-extensive with the existence of our knowledge. In other words, we *give* order to things, and anything outside the sphere of our control is precisely *out of control.* We are the ones, therefore, that *make* all the connections between things. More pointedly put, we can admit of having no relationship whatsoever to that which falls outside of the sphere of knowledge, for one cannot be connected to or dependent upon chaos.

Correlative to this notion of chaos is the belief that we must act only on the basis of what we know. But since we can only rely upon what we know, we must know everything in order to act with full confidence, reliably. Or we must portion out gradations of reliability with the methods of statistics and thereby regain our confidence through the mathematization of our hesitancy regarding what we know. Even our hesitancy becomes colonized with mathematics.

This leads to a sort of mania. We end up in a position of constantly trying to outrun the potential invasion of chaos into the order(s) we have made. We thus become consumptive of anything outside of the sphere of knowing, rushing to usher it into that sphere, because we have come to believe that it is only what we know that we can rely upon.

The notion of mystery, on the contrary, entails that the sphere of what we know is a limited one, but that which goes beyond that sphere contains implicate patterns and relationships which have their own integrity, which are not of our own making. Acting with propriety is acting on this assumption. More pointedly, the notion of mystery entails that we are connected to and dependent upon what falls outside of the sphere of knowing. On the basis of such a notion, we must act on the basis of ignorance. Our actions must become delicate and careful and attentive to what crackles beyond the boundaries that our knowledge has set. We must leave ourselves second chances to act again; we must act in ways that sustain the possibility of acting anew, acting differently (Berry 1983).

Mystery gives our attempts to *make* order a certain humility:

> The acquisition of knowledge always involves the revelation of ignorance. Our knowledge of the world instructs us first of all that the world is greater than our knowledge of it. To those who rejoice in abundance and intricacy, this is a source of joy. To those would-be solvers of "the human problem," who hope for knowledge equal to (capable of controlling) the world, it is a source of unremitting defeat and bewilderment. One thing we do know, and dare not forget, is that better solutions than ours have at times been made by people with much less information than we have. We know, too that the same information that in one [person's] hands will ruin the land, in another's will save and improve it (Berry 1987, 65).

It is so hard to imagine, given the proliferation of information in education, that better solutions than ours have, at times, been made by people with much less information than we have, and that, even with all our information and all our constructions and schemes, our actions could be ruinous if we remain within those constructions and "hear only [our] own words making up the world" (Le Guin 1987, 11).

A final thought. As mentioned in the beginnings of this paper, we are living out a logic that is centuries old, far older than the world of Descartes or Kant, and far more deeply embedded in our lived experience than is some abstract theory in an academic journal. This logic outlines how we have come to carry ourselves.

Gazing out the window to the pine and spruce trees, and this snow and the Chinook winds, it is almost impossible to realize that these things are not there *for me*. This is the sting of ecological insight, that these things have an agenda and an integrity of their own and they are not waiting for me to make sense of them or to graciously bestow them with order. However, once I deeply realize that these things are not here for me, they begin, so to speak, to turn away; they no longer pay any special attention to me; they are not formed up and tarted up for my perusal, like the false smiles of the colonized who have been ordered to put on a good front (to be presentable, like the roots of Kantian representation, formed to fit our demands), knowing that if they don't they will be made to mind. Suddenly, I am no longer a violent, interloping stranger that everyone has to keep their eye on and organize their actions around, the one who can make demands without attention to where I am, without attention to what is *already at work without me*. Suddenly, I belong here. I live here in relations of mutuality. The Earth becomes, not an object displayed according to forms of human understanding, but a home that embraces. I can become, as ecology suggests, deeply conversant with things, listening, asking, responding, withholding my actions, and acting with a sense of care for what I act upon, knowing that what I act upon will have an intimate say in how the conversation will go.

Such conversations are rare, and they are only possible insofar as they issue up out of a sense of the living integrity of my interlocutor and the belief that I and the one I engage are somehow mutually caught up in a living, vital relationship in which each of us needs the other to be what we are. We need each other, not just as raw, inert material upon which our forms can be forced, but as an alternate voice which puts into perspective my own en-forming activity, resists it, suggests alternatives, shows that my knowledge is not equal to the world but is just a possibility living among and sustained by others I have not constructed, and occasionally outright refuses such bestowals of meaning.

This is what conversation requires: that neither of us are interested in rendering this mutuality into a single voice of domination over the other, with all the images of colonialism that infuse threads of the logic we are living out, and the deep rage for imposing orderliness on anything that comes to meet us and the deeper hubris hidden in the mistrust that there may be orderliness without us. These are especially vital images for pedagogy to consider. Perhaps further considerations of what ecology is telling us will help slip pedagogy back into its element. In the end, ecology and pedagogy are deeply interwoven, to the extent that "no matter the distinctions we draw, the connections, the dependencies, remain. To damage the Earth is to damage your children" (Berry 1986, 57).

CHAPTER SEVEN

Student Teaching, Interpretation, and the Monstrous Child (1994)

Interpretive Readings of the Tales of Student Teaching and the Appearance of the Monster

> When I went to my practicum school the first day, I felt about this high, like a little kid again, going to school for the first time, beginning all over again, right from the beginning. And my cooperating teacher was, like, *my* teacher and I was a little kid in grade one. I was so afraid, but I knew I could get through it if I just kept going.

The liminal space indicated by the hyphenation "student-teacher" is a haunted and generative space, full of tales told to anyone who will listen. Those of us involved in the practicum experience often pass over a consideration of the deep mytho-poetic meaning of these tales in favor of interventions aimed at facilitating and easing this transition. Such well-intentioned efforts can all too easily be premised on the implicit assumption that there is no deep narrative-interpretive structure to these transitions, and that every difficulty confronted is somehow avoidable, that every pain points to a pathological condition—a dis-ease requiring a cure. From such a pathological/curative premise, student teaching can easily devolve into the belief that becoming a teacher would be simple and painless if only we could orchestrate it well enough. Perhaps even worse, we can mistakenly believe that simply making our student-teachers feel good is a sign of both their success and ours.

The interpretive effort of remythologizing pedagogy, found in the recent work of Jane White (1989) and Fay Head (1992) (see also Jardine and Field

1992), is aimed at restoring the phenomenon of student teaching and its existential transformations to their full, "original difficulty" (Caputo 1987). These papers begin to make this difficult liminal experience readable and understandable and decipherable as something more than simply an array of problems to be fixed. More strongly put, this work suggests that there are deep and irremediable difficulties inherent in the liminal space traversed by student-teachers *that cannot and should not be fixed.* Not all difficulties are a result of lack of effort or diligence or preparation or information, and understanding student teaching in such a pathological fashion not only foregoes a collective, communal understanding of its deep human meaning (with all of the ways in which such commonly held understanding can help to make the burden of those difficulties more bearable). Pathologizing student teaching can also subvert the transformative initiatory process itself (where one "becomes *another"* [Eliade 1975, 165]) by attempting to disassemble this often long and occasionally painful transformation of who one deeply *is* into the hurried accumulation of skills and techniques. And, if we take even a cursory look at the publishing industry related to education, pathologizing student teaching also makes these vulnerable "initiates" too easily prey to ever-new and easily purchaseable materials, kits, booklets, packets, texts, bells and gongs and whistles.   Inevitably, in such disassemblages, the difficulties of teaching become understandable as little more than technical problems requiring technical solutions.

The interpretive effort is directed squarely against this flattening, overly technical, surface reading of our lives as teacher educators and the lives of our students as initiates into the community of education. Interpretive work wishes to evoke and bespeak the figures that haunt us beneath the clean and literal surfaces of technique. It wishes to evoke the places where collective meaning resides — a haunted space where tales arc into tales and sense into sense, where the ambiguous passage of messages ensues, like the tale of the student-teacher cited above, rich as it is in its images of the passages between the young/student and the old/teacher, in its suggestions of the regenerative movement of the initiate back to the freshness and innocence and vulnerability of the child (Eliade 1968, 139) and in its hints as to how this movement regenerates the community with the initiates' infusion of "new blood" and the return of the community to is origins, what is originary/initial (Eliade 1968, 1975).

This telltale, generative "gap" between student and teacher (this "-") can thus be envisaged as a portal, full of opportunity (Hillman 1987), but also full of portend, warning: lessons to be learned. It is a gap between worlds and, in its lessons, it is a deeply pedagogic space.

Both Jane White and Fay Head allude to a wonderful and dangerous and often ignored figure found lurking in such gaps. This figure is full of portend. It both creates and appears in the gaps in the once-familiar world. But it does more than this: it *guards* those gaps, watchful, warning that life will be different if one "passes" through them. This is the figure of *the monster*. The figure of the monster that wanders through anthropological and mythological literature appears in the most unanticipated ways during practicum year when students are slipping through the liminal space between student and teacher. It is this figure that I wish to stalk in this paper.

In particular, I am interested in weaving an interpretive tale of how the child functions as a monster for a beginning teacher and how the de-monstrations of the child's questions can teach us a lesson in the very act that we teach them. But there is a second thread to this interest that is linked to the character of interpretation itself. The monstrous child is not simply one figure among many about which one might tell an interpretive tale. Interpretive work is *itself* haunted by the figure of the monstrous child in the guise of the young boy Hermes who, full of portend and livid de-monstrations, stands at the boundary or gap (Hillman 1987, 156)—this "-"—, both keeping this gap open and watching over it. Interpretive or hermeneutic work does not provide us with a clean and literal surface text. Its texts are full of haunted, generative gaps. Its readings of our lives are thus de-monstrative and exaggerated (Gadamer 1989) and thereby contain a certain irremediable risk and danger (one of the consequences of standing near the limen: "life erupts at the boundaries," [Devall and Sessions 1985] but such eruption can spell trouble).

In this way, an interpretive reading of the passages of student teaching (pun intended—both their liminal travails and the tales told *about* those travails) has a special twin relation to the monstrous child, linked to the images of generativity, natality, and renewal that are common to both interpretive activity and pedagogy itself.

The figure of the monstrous child thus provides a peculiar crossover between these two phenomena that has an effect on our understanding of teaching itself. The monstrous child (central as it is to both pedagogy and interpretation) has the effect of making the community of teaching understandable and visible to the initiate/student-teacher as something more than a flat, literal set of accumulable information and surface skills and techniques. Teaching becomes understandable as a deeply communal, *interpretive* act, constituted by "a consciousness that must leave the door ajar" (Hillman 1987, 154) (a door frequently *pulled* ajar by the demonstrative child). Teaching thus becomes understandable as a "community of conversation" (Gadamer 1983, 165) between the familiar, established world and the inevitable generativity and transformation of meaning that the entrance of the new, the young, the initiate/initial (the "new generation") portends.

Monstrousness and the "Factors" of Pedagogy

During rites of passage, features of the familiar world—the world wherein one felt at home and able (playing as this does with the wonderful etymological link between *habilite* and ability)—frequently become represented in the distended and disproportionate features of masks that are used by the elders to teach the initiates a lesson they will never forget (Turner 1987). The monstrousness of these depictions is due to the fact that some once-familiar feature of the world has been severed from its familiar place. These once-familiar features no longer fit where one would normally expect them to fit, nor do they arrive as normally expected. They thereby lose all sense of proportion. The initiate is thrown into a position of "ambiguity and paradox, a confusion of all customary categories" (7), because the initiate, too, feels out of place and loses a sense of proportion. The initiate no longer feels at home and no longer has the reliable, familiar guideposts in place to keep things in perspective (Jardine and Field 1992). Thus out of place, the commonplace becomes disproportionate, monstrous.

This liminal space (Turner 1987, 5–6) is a gap between worlds ("-") and as such, "the state of the ritual subject is ambiguous: [s/]he passes through a realm that has few or none of the attributes of the past or coming state" (5). However, what becomes visible in such an ambiguous gap—albeit in monstrous form—is precisely the deep and inviolable attributes of the community one is wont to enter. That is, "much of the grotesqueness and monstrosity may be seen to be aimed not so much at terrorizing or bemusing the neophytes as at making them vividly and rapidly aware of what may be called the 'factors' of their culture" (14).

Confronting such monsters and the demonstrations they provide is thus always done "with a pedagogic intent" (15). In the end, the initiates are "returned home" having, through their monstrous visions, come upon certain intractables that define and delimit the community they have entered. Having been at the limen of the community, they have come to understand its limits. Having passed *through* the limen of the community, they have had a momentary glimpse of the necessary openness of that limit to the new ones. In the community of teaching, guiding this return might bespeak the facilitating and curative work of student-teacher supervision. But still, certain "original difficulties" (certain "factors") simply *pertain* to the community that the new ones have entered. Thus, such facilitating and curative work is not a matter of *solving* these difficulties, but rather helping the new ones learn to live with them *well,* in a way that protects them and remains mindful of their original character. In the case of the community of teaching, this entails being mindful of the sustainable limits of the established world (what deeply pertains to it underneath the fads and fashions that come and go) but mindful, too, that that world must remain open to the new ones, the children.

Pedagogy and the Monstrous "Fact of Natality"

Aghast, we cover our faces, confused and unable to choose between expressions of disgust or nervous laughter. What a surprise . . . who could have imagined . . . such horror. There is a moment of black epiphany at the revelation of a particularly heinous crime—a moment that is both oracular and inexpressible. As we make out the shape of the crime, as we see it unfold like some putrid flower, one word sputters to our lips: "Monster." The choice of the word is instructive. Etymologically . . . it is related to *demonstrate* and to *remonstrate,* and ultimately comes from the Latin *monstrum,* an omen portending the will of the gods, which is itself linked to the verb *monere,* to warn. Monsters, therefore, were created to teach lessons. And they can still be pedagogical—even in an age that no longer believes in the gods or their messengers. Monsters need not look monstrous (Chua-Eoan 1991, 27).

Over the course of several years of practicum supervision, such moments of black epiphany have occurred with a wonderful regularity. Inevitably (it seems), as their first practicum experience approaches, students begin to slip into a confused, liminal state and inevitably there recurs in our practicum classes the outburst, expressed in loud, disproportionate horror: "What do I do if a child asks me a question?!"

This seems, on the face of it, to be a simple expression of anxiety and, of course, as a practicum supervisor, efforts ensue to help calm the fears—showing how the fear is common to many other students, how the anxiety arises because one is envisioning the child in the abstract before actually being in a particular classroom setting and so on. As Victor Turner (1987) suggests, the monstrousness is due in part to breakdowns in the familiar world common to those undergoing initiation.

But what of the demonstration? What does this monstrous vision of the child demonstrate? What is the pedagogical lesson buried in this black epiphany? Even though we might pathologize this phenomenon by linking it to its causes (e.g., the stress and anxiety of anticipating an upcoming practicum experience), interpretively speaking, causes are not necessarily commensurate with narrative meaning.

Indeed, with this example, it is clear that "the monster need not look monstrous" and that its effects and its images — its "telling" character — far outstrip its causes. What occurs is that a most ordinary and familiar feature of classroom life (a child asking a question) gets yanked out of its ordinary and familiar place and as such, it becomes a *monstrum:* the familiar world seems to break apart, to break open, and a lesson is foretold, an opportunity is voiced. However, such

opportunities are not plain, clean gifts; they trail dark and chaotic attachments to their unknown backgrounds, luring us further. One insight leads to another; one invention suggests another variation—more and more seems to pass through the hole, and more and more we find ourselves drawn out into a chaos of possibilities (Hillman 1987, 154).

This alluring portend of a "chaos of possibilities" and the disorienting vertigo experienced in being drawn out into this potent, ecstatic, generative swirl is one of the original difficulties that pertains to the community of teaching. It represents the generative, questioning, potent character of the young.

The monstrous character of the child's question is not an epistemic matter of failing to have adequate insider's knowledge or skills to respond (see White's [1989] emphasis on such knowledge as constituting the heart of student teaching as a rite of passage). Rather, this question demonstrates a fundamental factor of teaching that lurks beneath issues of established skills, knowledge, and carefully laid (lesson) plans: *the fact of natality* (Arendt 1969), and how, in spite of all of our concerted efforts to "teach them a lesson," the young simply keep coming (like opportunities, "more and more"), standing before us, ripping open ever anew what we have taken to be "established knowledge" and putting us and the world and the curriculum (our "course") into question again and again.

The profound speechlessness that is represented in this student-teacher's monstrous vision is thus the oracular speechlessness one has in the face of "life itself" (Smith 1988b, 175). Not coincidentally, the monsters that haunt anthropological and mythological literature "guarded the sacred spring and the spring of immortality (the Tree of Life, the Fountain of Youth, the Golden Apples)" (Eliade 1971, 165). "They guard every symbol . . . able to bestow *power, life,* or *omniscience"* (291). One can thus read contemporary efforts to place the child at the centre of curriculum as an unvoiced recognition of the centrality of generativity to the pedagogic enterprise. Not coincidental, either, is the ways in which the child has often been portrayed as wild ("all the children are wild" [Le Guin 1989, 47]) and animal-like (see Miller 1989). This monstrous vision is one of realizing a sudden animateness to the world beyond that for which I have planned and prepared. It is a vision of the world as populated by these kin of ours that live "beyond our wanting and willing" (Gadamer 1989, xviii). It is thus a vision of the child finally coming alive and providing, quite literally "when all is said and done," an uncontrollable obverse to my own well-laid plans.

In placing us into question, the child does not merely disturb what we know and cause our familiar world to disassemble and re-figure itself in response. *This monster wants "in."* In asking a question of us the child opens up this "us" and thereby *becomes one of us.* And this, in turn provides the community with a certain re-generativity and renewal. Buried beneath the established community and the lessons it wishes to teach the child is not only the *fact* of natality and generativity, but also the ways in which the community desperately *needs* this generativity of child's question for its own renewal.

The monstrous vision of natality thus courts a dark twin: an equally monstrous vision of our own mortality, the mortality of our world and our own deep need of the renewal that children demonstrate:

> We are always educating for a world that is or is becoming out of joint, for this is the basic human situation, in which the world is created by moral hands to serve mortals for a limited time as home. Because the world is made by mortals, it wears out; and because it continuously changes its inhabitants, it runs the risk of becoming as mortal as they. To preserve the world against the mortality of its creators and inhabitants it must be constantly set right anew. Our hope always hangs on the new which every generation brings (Arendt 1969, 192).

Or, linked more directly with anthropological and mythological tales of "primitive" cultures:

> Through its own duration, the World degenerates and wears out; this is why it must be symbolically re-created every year (Eliade 1968, 76).
>   It is a living world—inhabited and used by creatures of flesh and blood, subject to the law of becoming, of old age and death. Hence it requires a periodic renewing (45).

So, just as the child brings renewal, it also brings a vision of the *need* for renewal and therefore a vision of the mortality of the one who is asked the question. This is the existential (rather than epistemic) sense in which the child's question places this student-teacher in question. Her very being and becoming (and therefore ceasing to be) are evoked. This student-teacher's outburst did not express embarrassment (what one might feel in failing to *know* something). She was *mortified,* failing to *be* something. And this failure, this weakness, is *irremediable.* To learn to live well with this irremediable weakness means to understand and resolve herself to the fact that all her lessons and all her planning and all her effort and all her knowledge will come to nothing if those lessons cannot be reinvigorated and reenlivened by the young entering these matters and thus providing them with "new blood." And, despite all her wanting and willing, she cannot *make* a child ask a genuine question. She cannot *make* the child enliven the world (even though she may be able to set up conditions under which this is more likely, more possible, more invited). This just *happens* (Weinsheimer 1987) (or fails to happen), beyond our wanting and willing and in spite of the learned and measured manipulations we may use to provoke "children's questions" in a lesson. All we can do is deeply understand the world that the child is entering and then invite that child to enter. The child standing at the limen, desiring entrance, shows us our own irremediable limits, our own finitude.

This mortification signals the *vulnerability* of the community, its weakness. Its weakness is its openness and therefore its hope, its portend: "one's weakness bears one's future; one's inability is the place of one's potential" (Hillman 1987, 188). This weakness of the community of teaching is therefore also its strength, because the mortality of the world bespeaks our deep kinship with children and the generativity they represent (*gen* being the parallel Sanskrit root of *kin*). The weakness of the community of teaching, its vulnerability, is also its *openness to the young,* its "generosity" (one could playfully say, its "kind-ness"), its willingness and ability to take the point of view of the young in a way that protects the young and renews the world.

But there is another triple thread here: we also "*fear* the closeness of the gods in the myths which found [us]. Through them, we can be found out" (Hillman 1983, 42). There is something fearsome about the fact of natality and the wildness and animateness/animality it bespeaks (to say nothing of the mortality that these gods portend). Alice Miller's (1989) work explicates in horrible detail the consequences, not of the wildness and animality of the child, but of violent and ungenerative/ungenerous responses to that wildness. These violent responses are responses to something real, something genuine: the wildness and animality of the child is no joke and it threatens to overrun the established world as much as it portends to save it. De-monstration can go too far, things can get out of hand. Put differently and into the context of education, there is a danger lurking at the heart of "child-centered pedagogy." It is not just good news. It requires a wariness, a watchfulness that the headlong rush of the "chaos of possibilities" that the young bring does not despoil the living conditions under which the generativity of the young may be sustained and nurtured. The monstrous child reminds us that the world needs loving attention so that it may remain open and caring for the young. The monstrous child, in his or her very act of ripping open the limits of the community, reminds us that *there is* a limit/limen: certain factors *pertain* (but they will cease to pertain without the young entering them and making them new).

The second part of this thread is how this is also a profoundly *ecological* point. The monstrous, animate child represents the wild/wilderness (Le Guin 1989) and therewith foretells the dangers of our current ecological crisis. The fearsome response aimed at controlling the wild (which Le Guin also links up to the fearsome response aimed at controlling women) is unable to see how we *need* the wild/child for our own renewal: it is both our (re)source and our limit.

Thirdly, the wildness/monstrousness of the child is at once a demonstration of *my own wildness*—my own "child"/childishness/childlikeness. The monstrous child thus exposes my own gaps/vulnerabilities/ weaknesses/wounds. The child's question reopens the body (of established knowledge). The monster leaves a wound (Jardine 1992a; 1992b), a mark, and it works to *keep it open.* The monstrous child thus faces me with my own ability (or lack of ability) to remain open to the new, to face my own renewal: am I able, as a teacher, to listen anew to *this* child and let what I take to be established be reenlivened and made new through their questions? Again, this is the existential character of the transformations student-teachers undergo that transcends the mere epistemologies of "insider's knowledge." Despite all the curricular and developmental knowledge and all the teaching skills and techniques, it is always *I myself* who must take up the task of becoming a teacher, who must find the "life" in this knowledge, must risk finding its portals and gaps and entrances, and must find my own voice to speak it. *I myself* am the one who is directly culpable and directly at stake in this question of "becoming another." The monstrous child thus poses its question directly and inescapably to the generativity (or degenerativity) of my own life, my own tendency to have the established limits of my understanding of (more strongly put, my *living in*) the world calcify and harden—liminal gaps transformed into fixed methods or mindlessly repeated lessons or closed altogether; established wisdoms, senatorial when they invite the young, dried up into senilities which, in the end, despise the young (Hillman 1987); or the oft-told joke that "I had everything so well planned and then the damned kids showed up," funny, in its way, because it is so telling.

This ability or willingness to remain open to the world is a key element in the practice of teaching. But it has an important correlate. It requires not just *keeping myself open,* but also "keeping the world open" (Eliade 1968, 139). This monstrous child demands that *the world* must needs be open, porous, permeable, generative for the sake of its own renewal (and thereby for the sake of children). Through the child's question, we can thus catch a glimpse of the deep *interpretability of the world.* Only as interpretable (i.e., full of gaps and portals and passages) does the world become *enterable* for (and therefore renewable by) the child. Only in an interpretable world is pedagogy possible. Put differently, teaching is a profoundly interpretive activity.

So here is the final guise of the monstrous child that haunts us—a irresolvable convolution at the heart of teaching. The teacher stands at the point of the meeting of the young and the world and is responsible for both (Arendt 1969). The teacher must protect the usherance of the young into the world while at once protecting those factors that pertain to the world that make possible such usherance (open, porous limits which let in the young—*need* the young—but limits nonetheless). The teacher must lovingly and generously embrace this irresolvable difficulty, which simply convolutes again and again: for the entrance of the children will inevitably transform our precious belief that *we* always and already know what the limits are. Children add themselves to the world and the world becomes new, the limen shifts and flutters, however slightly. Teaching thus involves the luscious agony (Hillman 1983) of "dancing at the edge of the world" (Le Guin 1987), keeping the doors ajar, watchful.

Concluding Remarks

An interpretive reading of the telltales of student teaching aims, in its own way, to keep these tales open, telling. And the synchronicities between student teaching and rites of passage are simply too telling to avoid: a student-teacher told me just today how she has bought "a new outfit" for her first day, just like we all do if we can afford it, shedding the old and taking on new clothes, new (in)vestments. An old and wonderful story that is not just her story but also *ours* in this haunted, liminal space.

CHAPTER EIGHT

Wild Hearts, Silent Traces, and the Journeys of Lament (1993)

I

We live in skittering times, when the old reliables of our own invention are beginning to crack.

There is both sadness and adventure ahead, and there is pain to pay for the somnambulant beliefs in our own dominion.

Children sit in the middle of the maelstrom, full of belly giggles and little night tremor jolts, waiting for us to respond in kind.

Waiting.

"—have you forgotten?
All the children are wild" (Le Guin 1989, 47).

The old pedagogies will no longer do in the face of the ecologies of flesh and bone and breath and the Earth's dear heart, torn.

"you lived there
—have you forgotten?" (47)

What have we forgotten is that, below the distended head we carry like a crown, we live there still, we live there now, entrails wound around trees and down into Earth's moist substance.

An open wound we bear in the belly "—have you forgotten?" (47)

These children are some of us and we are some of them, wild, full of the ecologies of flesh and bone and breath,

And air filled with ancient smells and smoke spirals on the hills.

All around us is the urgency, borne up out of the dark gut and assaulting the head which we came to believe in as a fortress against death, against touching winds, against those little unmarked joys that the whiff of cedar brings without expectation.

That luxurious isolation is over.

Our words bleed out into the Earth's fleshy countenance.

The open wound that is our fleshy measured life.

"Relentless clock of meat" (Ginsberg 1984)

Healed over only at great casualty.

And we find that we live on an Earth which will take away our breath as surely as it now bestows it.

II

Old voices in old keys in old harmonies are beginning to sing again, composing and decomposing without our notice, echoing within.

Composing and decomposing.

The chest becomes a dark chamber and the breath a black wind dancing out into the Earth's embrace.

Frosting up into winter air, white, and sucked back in to moist tubular shadows.

Our essays are borne up and sung in throats and written in blood and etched on skin, then fading, falling, making soils rich and fertile again.

The pedagogies of precious returning.

The curriculum full again of courses running as they must.

As they always have unnoticed for so long "—have you forgotten" (Le Guin 1989, 47)?

"Rainwater moves from mystery through pattern back into mystery" (Berry 1987, 4).

"We are in the midst of it, and it is in the midst of us, for it beats in our very blood, whether we want it to or not" (Merton 1972, 297).

Water trickled into valley and crevices, hidden from our search for givens that would place all things before us.

Springing forth again the inevitabilities that lace our lives to Earthlife whole and hale.

Our lives are instances of this Earthlife, occasions where the whole Earth breathes out and speaks.

We are instance not exception.

We are not exceptional, for the Grey Jay swoop for scraps, and the little wet watermelon peel limping to decay on the compost heaped to wasp and flybuzz squirm of returning are also articulations of the whole.

Still warm horseshit piled on the compost erupting with peas and potatoes, vague miracles.

More articulate, more reliable, more full of rich integrity than the mumbled numbness of our lucid charts and maps.

"Stubborn particulars of grace" (Wallace 1987) where the whole is ushered *"here* and *here* and *here"* (Wallace 1987, 111).

Where the whole is held here in harmonies of kinship *here,* with *this one.*

And like whispers to like in ways we did not invent.

Ways which hold us up and measure the echo of our footfall, again in harmonies unnoticed.

These children are some of us and we are some of them, wild kin, full of the ecologies of flesh and bone and breath and air.

"An order of memories preserved in songs and stories, in *ways"* (Berry 1983, 73).

This essay is an Earthly event, full of dark soils and the long neck quivers of ducks finding south again as they must.

This is the balancing act of a phenomenology: to retain the exquisiteness of *this* while at once cracking it open, eggyolk yellow sunlight breakfast winterwindows crackled with ice.

This is the balancing act of a pedagogy: to retain the exquisiteness of this child's life while at once cracking it open into the textures of the Earth.

Signs of spring whiffed in the sun now arcing higher.

III

Old green ears are growing again on the sides of hills, in the drips of rainwater moving from mystery to pattern to mystery.

Our hearts somehow know that they are growing *again*.

Something is being stirred in this skitter and slip.

And it doesn't come from before but from beneath, belying the swoon of memory and the romantic fantasies for a childhood none of us ever really lived.

We encase wild hearts that know of their beating if knowing is placed back where it belongs, in the meat between the ribs and the pull downwards of faint gravities.

"—have we forgotten?" (Le Guin 1989, 47)

This is not longing for long ago, but longing for times now that are old, long-standing.

A longing and a grieving, "opened out" (Bly 1990, 11).

"Grief is not a permanent state; it is a room with a door on the other wall" (Bly 1990, 11).

Going down into the flesh wound that droops below the head, wild at heart and silent.

"Wounds need to be expanded into air, lifted up on ideas our ancestors knew, so that the wound ascends through the roof of our parents' house, and we suddenly see how our wound (seemingly so private) fits . . ." (Bly 1990, 12).

But it is too easy to be caught in the Enlightenment metaphors of height, ascendancy, and light.

Ascendants which *let go* of the entrail wounds and spiral out of sight of the Earth and our fleshy inheritances.

Developmental theories always proceed *up* to higher and higher stages.

(Daddy is tall and strong and has dominion over all things.)

"Peeling off layers of flesh as they go, cleaving the abdomen and climbing up into the head" (Le Guin 1987).

We grow *up*.

We imagine an Earth left behind and a fleshless Epistemic Ego.

Puffed full of Teutonic accuracy and a mathematics cleaned of its contingencies, life gutted like a trout, slit from chin to belly, leaving the head intact, eye blinking, lines and that therefore it is.

But consider.

Ecology doesn't move us *back* to a time before, but *down* to a place filled with darkness *now*.

The flesh wound that droops below the blinking head, wild at heart and silently out among the Earth's ways already, breath and bone and air.

Portals half open, half lit, half noticed.

"Opportunities are not plain, clean gifts; they trail dark and chaotic attachments to their unknown backgrounds, luring us further" (Hillman 1987, 54).

The underbelly shadows that lurk our lucid givens, plain and clean.

The taking-away inevitably that comes with our gifts.

The losses laced to gains, coming and going, "A swinging gate" (Suzuki 1986).

Suggestions, possibilities, provocations, hints, and hopes and glancing blows luring our speech out into the open.

Throats filled with aspiring breath plumed out into the winter air.

The forgotten and frightened asthmatic little boy that giggles under the guides of maturity, vicious now in his wildness, breath denied, aspirations barely understood.

It will take time to forgive myself—I hid and did the right thing in such hiding—and to find the fierceness hidden in the hostility.

It will take time to forgive myself and outlive the hidden fears that lurk in the moist eyecorners.

Breath halts.

Stalking the woods, a deer's antler arcs up out of the snow, framing deadrustle grasses and red berries.

IV

Buried under the weighty heads we carry there is an Earth going on without our re-signing.

Buried under the weighty enclosure of our own words and signs and significations there is a going-out into the Earth's ways.

Embodied ways, fleshy ways, the bounce and bump and muddy squishes that we educate children out of, teaching them as we do to climb up into their heads and join our frightened numbers, our sad enumerations.

The Earth becomes mathematized and things don't quite add up any more.

Our living becomes vaguely incomprehensible.

But once the words and signs and significations begin to crack open and slip and shift, the choices become clear.

And postmodernism (that ugly designation for what comes after our heady confidences) becomes either the beginning or the end.

The difference is so important that our lives depend on it in ways that we can't name.

It may be ushering the totalization of signs.

The totalization of urban(e) theorizing done by those who have nothing else to do, who have no real work to plant them in a place—*here,* not just anywhere—except skittered sign play and who live in places, *constructed* of skittered signs.

Where sign refers to sign refers to sign refers to sign and we become wrapped up senseless in this soft cocoon of words.

This soft cocoon of words.

Hanging by a thin ligament somewhere, precariously near the mucous lips of a predator threading its way along the thin branches.

Chomp.

This makes postmodernism the consummation (chomp) of the long history of representation.

Our Earthly flesh becomes pictures out in front of us, readied and steadied for pornographic delightenment and manipulated interpenetrations and the educated giggle of drunken literati who tongue-cluck at the ways of the hand and the breath and the heart.

Accusations of feigned naiveties.

And speech finds its consummation (chomp) in advertising and the semantic twists of political charm.

Speech with no depth that can no longer wind its way into wounds that open out beyond their designations.

Speech which is not only a wafer-thin surface but which elevates the plays of such surfaces to the status of literacy.

Signs that have lost their point, their prick.

The male with no fierceness, caged by the confusion of hostility with wildness.

And these important surface-plays of presentabilities are finally wrapped back around themselves like a winding-cloth, swaddling the infant from movement or shrouding the unmoving dead.

Laughs begin as the air in this air-locked vault shallows, unnoticeable at first.

Muffling grins begin to pus.

Postmodernism as a last gasp, hopeless.

Not the portend of a last word, but the ugly truth that now we can *never* stop talking, that all we have is the world of words that we have made.

"A world made of words" (Le Guin 1987, 12).

That our lives are stories and no longer that which the story might be *about* and might fail to fully say.

Nothing left to hold us back from designing all things.

Full announcement, self-possessed: our ascendancy is complete.

And at that moment, the nightbirds turn away and scatter,

The eagle's winding arc widens, soon to be unseen.

The frightened deer no longer venture to this pool to drink deep draughts of cold water.

Ecology reminds us that if *this* is what postmodernism is (mis?) understood to be, it is a sign of impending death.

V

Or perhaps it is the totalization of signs to the point where the pus-y facade *cracks* open.

Perhaps it is the totalization of signs to the point where the *difference* between the surface signs and the hidden depths becomes unavoidable in our signing.

We must now talk about the fact that our lives are different than our words, that our children are different than what we say, but that, in the midst of such difference, kindness ensues, and like speaks to like without our earnest intervention.

And to talk of such Earthly belonging, our talk becomes hesitant, stuttered, skittery, single-sentenced.

Like this.

And "whatever remains unsaid in us is forever angling to come into view" (Hillman 1983, xv).

Caught and glimpsed but not captured.

"Poetry is a tool, a net or trap to catch and present; a sharp edge; a medicine, or a little awl that unties knots" (Gary Snyder, cited in Berry 1983, 28).

It squirms in our nets.

It squirms in our nets.

Yellowslit eyes and a heartbeat you can taste in the air, surge electric nicklespittle fear.

Postmodernism as a sign of life bubbling under the signs.

And this paradox is good news, something we must learn to live with.

It is a herald, a clang, a bell, an echo.

And it sits in the belly, not the head, below the wheeze of discursion.

If this reading is the right one, postmodernism is inhabited by a vital pedagogic impulse.

That under the burgeoning edifices of educational theory and the slick tricks of practice and the earnest reflectiveness of the practitioner, children are giggling and dancing out into the flesh of the Earth's ways.

That under the edifices may be forms of ecological disaster wrought on the wild heart of children.

Deeper hints that our edifices may be forms of ecological disaster wrought on the wild heart of children.

Wrung into the wild heart of the Earth.

Words and effort and energy wasted.

There are 107 developmental strands to the proposed new Language Arts Curriculum Guide.

Think of what is required to sustain such a beast,

to feed it,

to keep track of it, to dispense of its waste.

To move its taming bulk into the wild areas of language and the wild hearts of children.

Think of how convinced we must be that the edifice of signs will suffice if only we are effortful and artful enough in our accuracies.

We all feel the urgencies all around us.

Things beginning to slip and shift.

And we run faster as a result, multiplying our divisions in the sad Enlightenment hope of outrunning our fleshy inheritance, needing to keep track of every meticulous articulation and subdivision of our own unsustainable invention.

Losing our love of language and our sense for its aromas, our ears for its harmonies.

It may be that this life bubbling under the signs needs not a word from us, not a word.

That like might *already* speak to like without our earnest intervention, needing only our gentle attention, mindful to the ways that whisper within and without.

In the little sidewalk crevice dirt, a dandelion whispers up and burps a million seeds, unseen.

We need not feel guilt for the Enlightenment parade and the measured steps we ascended.

That silent, gentle force that turns plants to light is the silent, gentle force that dragged us up into our heads.

What we need is lament and mourning and grieving for what we have done in the name of our own ascendancies and aspirations.

We need loud wails and drums beaten and quiet halts in the midst of things and prayers shouted out into the exhaustible airblue arch.

We need remembering of our love of the Earth signed in the breath and beating in the blood whether we know it or not.

VI

Ecology keeps whispering in our ears that the sealing off of signs from things is the sealing off of things from signs, so that signs are no longer readable as signs of anything else but further signs.

Notice that these dragonflies *hover* and they can halt the inward curvatures of sun, and round the sky, waiting, ancient things, buzzed of swamp and moisture and Earthy smells, *insects,* built in ligamented sections, hard, brittle, flat black and bright green, this one, hovering.

Its hovering is a sign that there are flowers nearby.

But it will finally *light* somewhere in recognition that it does not contain its life in its hovered signing, but relies for its life on the moment of intersection.

Leaving behind faint pollen traces that belong there, borne to new life, unable without this intersection.

Beyond the signs to signs to signs to signs there is *already* an Earth which works perfectly well without our tracing words and in whose harmonies we already live.

One forest turns to winter and the heartache can be felt only in the traces of the work done there, in the traces of the breath spirals exhausted there in gathering wood for the cold's delight, in the marks on the hands and in the heart.

Wood gathered without a word, leaving traces that return, gentle intersections.

VII

We live in skittering times.
The gifts may not arrive.
"—have you forgotten?
All the children are Wild." (Le Guin 1989, 47)

"Good Wild Sacred" (Snyder 1984)

CHAPTER NINE

American Dippers and
Alberta Winter Strawberries (1997)

I

In the seventh month the Fire-star declines,
In the ninth month winter garments are handed out.
The eleventh month comes with the blustering wind;
The twelfth month, with the shivering cold.
Without cloak or serge
How are we to see the year out?

from "In the Seventh Month," compiled in the *Shih Ching,* Seventh to Twelfth century B.C., China (in Lui and Lo, eds., 1990, 90)

The American Dipper is a small black bird, half-way in size between a robin and a sparrow, ℧-shaped with stubby upright tail and head balanced high.

These Dippers are common all year long along the Elbow River that winds out of the Rocky Mountains and through Calgary, Alberta. Walked this past winter, in −40 degree winds, along the Elbow and its swirls of ice fog over rare still patches of open water, most of the river steeply hurrying east.

Dippers. Swim underwater upstream about ten feet at a go, feeding on water-carried food. Then standing there dipping up and down while waiting for the next dive. Or rush to low water-surface flights full of a distinctive twittery warble. Like the muskrat in the steaming cold beaver pond nearby, remaining here as I leave. Remaining here, −40, what little left low-riding sun setting.

Leaving, under the darkening air-blue arch of what I hope is a gathering Chinook, bitter cold that breaks your bones.

Home, eating freshly bought fresh strawberries. Delicious red juicydrip taste. Then suddenly grotesquely beautiful. Suddenly out of place. And as these strawberries begin to taste become unbecoming of this place and this cold, I end up feeling out of place as well. Eating these strawberries betrays something of those Dippers and that ice and my living here.

II

In the seventh month the Fire-star declines,
In the ninth month winter garments are handed out.
Spring days bring us the sun's warmth,
And the orioles sing. (1990, 9)

I recall growing up in southern Ontario, in what was then a small village crouched between Hamilton and Toronto, just at the west end of Lake Ontario—full of black-orange orioles, singing, and their droopnests branchended on silver maples or Royal Oaks, named like the red-and-white dairy trucks that delivered milk and cheese and butter and eggs.

Undeniable ecological memory, stored too deep, it seems, to switch. Earthy flesh memory born(e) in the body of the child I was raised. We carry memories of where we were born, and the triggers of such memories are themselves bodily. So "bioregions" are not simply places with objectively nameable characteristics. They infest our blood and bones, and become odd, unexpected templates of how we carry ourselves, what we remember of the Earth, and how light and delicate are our footsteps in all the places we walk.

Burlington was, in the 1950s, a market garden area—small forty- to sixty-acre farms bursting full—and an area with many canning factories and wooden fruit-basket factories (up in Freeman, a once-named crossroad near the rail lines that has since disappeared). Back when such things mattered, Burlington was right at the hinge between the north/south rail route up from the fruit-growing areas of Lincoln County (Vineland, Jordan Harbour, east to Niagara-on-the-Lake) and the east/west rail routes to Toronto, Kingston, and on to Montreal, or west to London and Sarnia. Swimming, 1957, Lake Ontario, off the redbrick Legion Hall parking lot south of Water Street, just after the Aylmer Factory had burped out the leftover bilge of tomato canning, and how this hot-scented red-scummed flotation that made the water flapthick muffled and fly surface buzzy melding clear into the Polio Scares and the summers of no swimming at all.

I remember, growing up in Burlington—then a small village of 6,000, since overgrown into the bedroom of nearby city condensations—having to *wait* for strawberries.

Their appearance once *meant* something deeper and more difficult than their obvious bright pleasure-presence to the tongue: about place, about seasonality, about expectation, about era, about arrival, about remembering, about reliance, about resignation and hope, about time and its cyclicalities.

Strawberries once *belonged* somewhere. They thus arrived, not as objects but as bright and brief heralds.

III

In the sixth month we eat wild plums and grapes,
In the seventh month we cook sunflower and lentils.
In the eighth month we strip leaves of their dates,
In the tenth month we bring home the harvested rice,
We make it into spring wine
For the nourishment of the old.
In the seventh month, we eat melons. (1990, 11)

Pulling strawberries into continuous presence, into continuous, indiscriminate availability is, in its own way, a sort of objectivism.

These Alberta winter strawberries are only in the most odd of senses *here* in my hands, even though, clearly, *there they are.* Something about eating them is potentially dangerously distracting. They are no longer exactly Earth produce. They are commodities lifted off the Earth and floating above it, taking me with them. As they begin to float up into detached commodification alone, I, too, begin to float, detached, unEarthly.

Such odd, objective strawberries, ripped out of the Earthy contexts of their arrival—no Earth to smell, no resignation to waiting fulfilled, no sunny warmth—can, however, also be alerting. This subtle disruption of a sense of seasonality that transports and technologies have brought us: odd pleasures, since, in the grip of cabin fevers, these strawberries have also saved my life.

IV

In the fourth month the small grass sprouts,
In the fifth month the cicadas sing.
In the eighth month we harvest the field,
In the tenth month the leaves begin to fall.
The eleventh month comes with the hunting of badgers;
Taking those foxes and wildcats,
We make into fur coats for our lord. (1990, 10)

I am living in a place that a hunter should live. It is a harsh place, where the summer will yield potatoes and peas and not much else.

Huge moose cow, great Alert Being, chest deep in snow near the river. An odd coalescing point in a large habitat. This moose body as a place of great intensity and great need. Munching on those small fir-tree tips seemed ludicrous and courageous and near-impossible all at once.

I cannot quite sit down and make my way here. Too many things, of necessity, must be like winter strawberries, like me, unable exactly to *live* here, even though *here we are.*

I can't stop remembering a tomato tossed out against the backyard fence in late Ontario fall and yielding, of itself, without me, beyond desire and necessity, a great green clump of bushes four feet across, with that unmistakable near-acrid smell of fat and furry vines.

Tomatoes as places of great intensity, great Alert Beings whose absence I grieve. Ecological grieving for that deeply imprinted place where I was raised.

Ecological grieving for the waiting that is no longer necessary.

V

In the fifth month the locust stirs its legs,
In the sixth month the grasshopper vibrates its wings.
In the seventh month, out in the fields;
In the eighth month, about the doors.
In the tenth month the crickets
Get under our beds. (1990, 10)

Winter has cracked, or at least blinked. I, too, am a Great Alert Being, surprised to find that some of that alertness, as well as some grieving, is carried here, to Alberta, from the place I grew up.

It took me years to even begin to actually experience this place and its beauties. Heartbreaking blue against the yellow green of pines.

Hale-Bopp's nightly bristle, Mars between Leo's legs, Orion setting.

And the Fire-Star—Antares, Great Red Giant in Scorpius—soon to rise.

REFERENCES

Aoki, T. 1987. "In Receiving, a Giving: A Response to Panelists Gifts." *Journal of Curriculum Theorizing, 7.*

Arendt, H. 1969. *Between Past and Future: Eight Exercises in Political Thought.* New York: Penguin Books.

Aristotle. 3rd C. B.C.E., reprint 1941. *The Basic Writings of Aristotle.* R. McKeon, ed., New York: Random House.

Berman, M. 1984. *The reenchantment of the world.* Toronto: Bantam Books.

Berry, T. 1988. *The Dream of the Earth.* San Francisco: Sierra Club Books.

Berry, W. 1983. *Standing by words.* Berkeley: North Point Press.

Berry, W. 1986. *The Unsettling of America.* San Francisco: Sierra Club Books.

Berry, W. 1987. "A Letter to Wes Jackson." In *Home Economics.* Berkeley: North Point Press.

Bly, R. 1990. *When a Hair Turns to Gold.* St. Paul: Ally Press.

Bordo, S. 1987. *The Flight to Objectivity.* Albany: State University of New York Press.

Caputo, J. 1987. *Radical Hermeneutics: Repetition, Deconstruction and the Hermeneutic Project.* Bloomington: Indiana University Press.

Chua-Eoan, H. 1991. "The Uses of Monsters." *Time* (August 12): 27.

Descartes, R. 1640, reprint 1955. *Selections.* New York: Charles Scribners' Sons.

Devall, B. and Sessions G. 1985. *Deep Ecology.* Salt Lake City: Peregrine Books.

Eliade, M. 1968. *Myth and Reality.* New York: Harper and Row.

Eliade, M. 1971. *Patterns in comparative religion.* New York: Meridian Books.

Eliade, M. 1975. *The Quest: History and Meaning in Religion.* Chicago: University of Chicago Press.

Fox, M. 1983. *Original Blessing.* Santa Fe: Bear and Company.

Gadamer, H. G. 1977. *Philosophical Hermeneutics.* Berkeley: University of California Press.

Gadamer, H. G. 1983. *Reason in the Age of Science.* Cambridge: M.I.T. Press.

Gadamer, H. G. 1989. *Truth and Method.* 2nd revised edition. New York: Crossroad Books.

Gick, M. and K. Holyoak. 1983. "Schema Induction and Analogical Transfer." *Psychology,* 15: 1–2.

Ginsberg, A. 1984. *Collected Poems.* New York: Harper and Row.

Habermas, J. 1972. *Knowledge and Human Interests.* Boston: Beacon Books.

Hahn, T. N. 1986. *The Miracle of Mindfulness.* Berkeley: Parallax Press.

Hahn, T. N. 1988. *The Sun My Heart.* Berkeley: Parallax Press.

Head, F. 1992. "Student Teaching as Initiation into the Teaching Profession." *Anthropology and Education Quarterly,* 23: 89–107.

Heidegger, M. 1954, 1968. *What is Called Thinking?* New York: Harper and Row.

Heidegger, M. 1957, 1971a. *On the way to language.* New York: Harper and Row.

Heidegger, M. c.1950, 1971b. *The Question Concerning Technology.* New York: Harper and Row.

Heidegger, M. 1962, 1972. *Time and Being.* New York: Harper and Row.

Heidegger, M. 1977. *Basic writings.* New York: Harper and Row.

Heidegger, M. 1928, 1978. *The Metaphysical Foundations of Logic.* Bloomington: Indiana University Press.

Heidegger, M. 1925, 1985. *The history of the concept of time.* Bloomington: Indiana University Press.

Heidegger, M. 1936–46, 1987. *The end of philosophy.* New York: Harper and Row.

Hillman, J. 1983. *Healing Fiction.* Barrytown: Station Hill Press.

Hillman, J. 1987. *Puer Papers.* Dallas: Spring Publications.

Husserl, E. 1902, 1970. *Logical Investigations.* London: Routledge & Kegan Paul Ltd.

Husserl, E. c.1925, 1960. "Phenomenology and Anthropology," In *Realism and the Background of Phenomenology,* ed. R. M. Chisholm. Illinois: Free Press, 154–70.

Husserl, E. 1913, reprint 1969a. *Ideas.* New York: Humanities Press.

Husserl, E. 1910–11, 1969b. *Formal and Transcendental Logic.* The Hague: Martinus Nijhoff.

Husserl, E. 1932–34, reprint 1970a. *The Crisis of European Science and Transcendental Phenomenology.* Evanston: Northwestern University Press.

Husserl, E. 1930, 1970b. *Cartesian Meditations.* The Hague: Martinus Nijhoff.

Jardine, D. W. 1990. "Awakening from Descartes's Nightmare: On the Love of Ambiguity in Phenomenological Approaches to Education." *Studies in Philosophy and Education,* 10: 211–32.

Jardine, D. W. 1992a. *Speaking with a Boneless Tongue.* Bragg Creek: Makyo Press.

Jardine, D. W. 1992b. The Pedagogic Wound and the Pathologies of Doubt. *Proceedings of the Second International Invitational Pedagogy Conference,* ed. B. Levering, M. Van Manen, and S. Meidema.

Jardine, D. and Field, J. 1992. "Disproportion, Monstrousness and Mystery": Ecological and Ethical Reflection on the Initiation of Student-Teachers into the Community of Teaching. *Teaching and Teacher Education,* 8: 301–10.

Jardine, D. and Morgan G. A. V. 1988. "Analogy as a Model for the Development of Representational Abilities in Children." *Educational Theory,* 37.

Kant, I. 1787, reprint 1964. *Critique of Pure Reason.* London: Macmillan.

Kant, I. 1784, reprint 1983. What is enlightenment? In *Perpetual peace and other essays.* Indianapolis: Hackett Pub. Co.

Kermode, F. 1979. *The Genesis of Secrecy.* Cambridge: Harvard University Press.

LeGuin, U. 1987. *Buffalo Gals and Other Animal Presences.* Santa Barbara: Capra Press.

LeGuin, U. 1989. Women/wilderness. In *Healing the Wounds: The Promise of Ecofeminism,* ed. J. Plant. Toronto: Between the Lines Press, 45–47.

Lui, W. and Lo, I., eds. 1990. *Sunflower Splendor: Three Thousand Years of Chinese Poetry.* Bloomington: Indiana University Press.

MacDonald, J. 1975. "Curriculum and Human Interests", (203–29). In *Curriculum Theorizing: The Reconceptualists,* ed. W. Pinar. Berkeley: McCutchan Publishing Corporation.

Merleau-Ponty, M. 1970a. *Signs.* Evanston: Northwestern University Press.

Merleau-Ponty, M. 1970b. *Phenomenology of Perception.* New York: Humanities Press.

Merton, T. 1972. *New Seeds of Contemplation.* New York: New Directions Books.

Miller, A. 1989. *For Your Own Good: The Hidden Cruelty in Child Rearing and the Roots of Violence.* Toronto: Collins.

Nandy, A. 1987. *Traditions, tyranny and utopias.* Delhi: Oxford.

Nishitani, K. 1982. *Religion and Nothingness.* Berkeley: University of California Press.

Norris-Clarke, W. 1976. "Analogy and the Meaningfulness of Language about God: A Reply to Kai Nielsen." *The Thomist,* 40: 176–98.

Phenix, P. 1975. Transcendence and Curriculum. In *Curriculum Theorizing: The Reconceptualists,* ed. W. Pinar. Berkeley: McCutchan.

Piaget, J. 1952. *Origins of Intelligence in Children.* New York: International Universities Press.

Piaget, J. 1968, 1970. *Structuralism.* New York: Harper and Row.

Piaget, J. 1965, 1971a. *Insights and Illusions of Philosophy.* New York: Meridian Books.

Piaget, J. 1954, 1971b. *The Construction of Reality in the Child.* New York: Ballantine Books.

Piaget, J. 1942, 1973. *The Psychology of Intelligence.* Totowa: Littlefield, Adams and Co.

Ricouer, P. 1970. *Freud and Philosophy.* New Haven: Yale University Press.

Sartre, J. P. 1970. "Intentionality: A Fundamental Idea in Husserl's Phenomenology." *Journal for the British Society for Phenomenology,* 1, 3–5.

Schopenhauer, A. 1966. *The World as Will and Representation.* vol. 1. New York: Dover Books.

Smith, D. 1988a "From Logocentrism to Rhysomatics: Working Through the Boundary Police to a New Love." Paper presented at the Bergamo Conference on Curriculum Theory and Classroom Practice, Dayton, Ohio, October 1988.

Smith, D. 1988b. "Children and the Gods of War." *Journal of Educational Thought,* 22, 173–77.

Smith, D. G. 1988c. "On Being Critical about Language: The Critical Theory Tradition and Implications for Language Education." *Reading-Canada-Lecture,* 6.

Smith, D. G. 1988d. The Problem of the South is the North (but the Problem of the North is the North). *Forum of the World Council on Curriculum and Instruction,* 2.

Smith, D. G. 1991. "The Hermeneutic Imagination and the Pedagogic Text. In *Forms of Curriculum Inquiry,* ed. E. Short. Albany: SUNY Press.

Snyder, G. 1980. *The Real Work.* New York: Wm. Scott McLean.

Snyder, G. 1984. *Good Wild Sacred.* England: Five Seasons Press.

Suzuki, S. 1986. *Zen Mind, Beginner's Mind.* New York: John Weatherhill, Inc.

Turner, V. 1987. "Betwixt and Between: The Liminal Period in Rites of Passage." *Betwixt and Between: Patterns of Masculine and Feminine Initiation,* LaSalle, Illinois: Court Publishing Company, 3-19.

Wallace, B. 1987. *The Stubborn Particulars of Grace.* Toronto: McClelland and Stewart.

Weatherford, J. 1988. *Indian Givers: How the Indians of the Americas Transformed the World.* New York: Fawcett Columbine.

Weinsheimer, J. 1987. *Gadamer's Hermeneutics.* New Haven: Yale University Press.

White, J. 1989. Student Teaching as a Rite of Passage. *Anthropology and Education Quarterly,* 20:177–95.

Wittgenstein, L. 1968. *Philosophical Investigations.* Oxford: Basil Blackwell's.

Studies in the Postmodern Theory of Education

General Editors
Joe L. Kincheloe & Shirley R. Steinberg

Counterpoints publishes the most compelling and imaginative books being written in education today. Grounded on the theoretical advances in criticalism, feminism and postmodernism in the last two decades of the twentieth century, Counterpoints engages the meaning of these innovations in various forms of educational expression. Committed to the proposition that theoretical literature should be accessible to a variety of audiences, the series insists that its authors avoid esoteric and jargonistic languages that transform educational scholarship into an elite discourse for the initiated. Scholarly work matters only to the degree it affects consciousness and practice at multiple sites. Counterpoints' editorial policy is based on these principles and the ability of scholars to break new ground, to open new conversations, to go where educators have never gone before.

For additional information about this series or for the submission of manuscripts, please contact:

> Joe L. Kincheloe & Shirley R. Steinberg
> 637 West Foster Avenue
> State College, PA 16801